Power Conversion and Electrical Machinery

Power Conversion and Electrical Machinery

Elizabeth Turner

NY RESEARCH PRESS
New York

Published by NY Research Press
118-35 Queens Blvd., Suite 400,
Forest Hills, NY 11375, USA
www.nyresearchpress.com

Power Conversion and Electrical Machinery
Elizabeth Turner

International Standard Book Number: 978-1-64725-432-2 (Hardback)

Cataloging-in-Publication Data

Power conversion and electrical machinery / Elizabeth Turner.
p. cm.
Includes bibliographical references and index.
ISBN 978-1-64725-432-2
1. Energy conversion. 2. Electric machinery. 3. Power electronics. 4. Power semiconductors.
5. Electric current converters. 6. Electromechanical devices. I. Turner, Elizabeth.
TK2896 .P69 2023
621.042--dc23

Contents

Preface

Electrical machinery and power converters are the two main parts of any electrical energy conversion system. Electrical machines are used for achieving power conversion of electrical energy into mechanical energy and vice versa. Power conversion is done with the help of a power converter. The primary function of a power converter is to convert alternating current (AC) into direct current (DC) and vice versa. It is also used to change the voltage or frequency of the current. Electrical machines use electrical energy to function. They convert mechanical energy to electrical energy or vice versa. Motors, transformers, and generators are the three primary categories of electrical machines. An electric vehicle is a type of electrical machine in which a power converter converts electrical energy stored in batteries to mechanical energy that propels the electric vehicle. This book outlines the applications of power conversion and electrical machinery in detail. Coherent flow of topics, student-friendly language and extensive use of examples make it an invaluable source of knowledge.

The information contained in this book is the result of intensive hard work done by researchers in this field. All due efforts have been made to make this book serve as a complete guiding source for students and researchers. The topics in this book have been comprehensively explained to help readers understand the growing trends in the field.

I would like to thank the entire group of writers who made sincere efforts in this book and my family who supported me in my efforts of working on this book. I take this opportunity to thank all those who have been a guiding force throughout my life.

Elizabeth Turner

1

Basic Concepts of Modeling

1.1 Basic Two-pole Machine Representation of Commutated Machines

The size of the DC machine depends on the main or leading dimensions of the machine namely armature core length L and diameter of the armature D. As the output increases, the main dimensions of the machine D and L also increases.

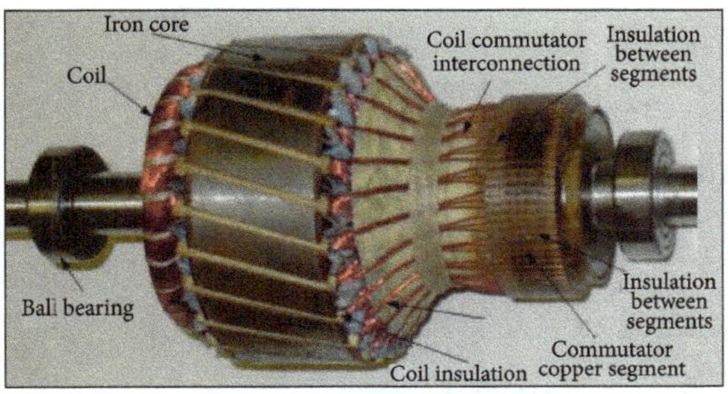

(a) Armature of DC machine.

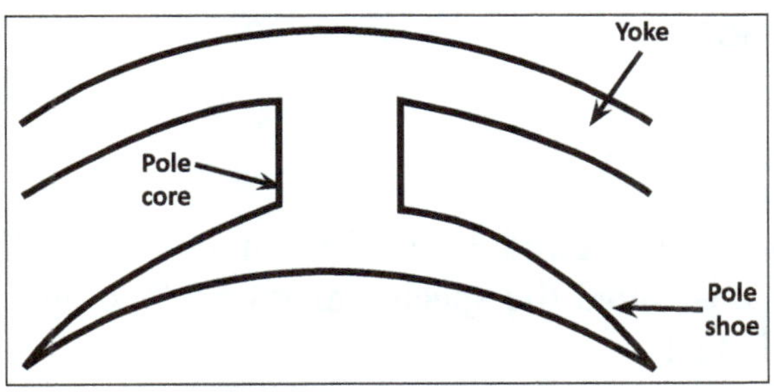

(b) Pole and yoke of DC machine.

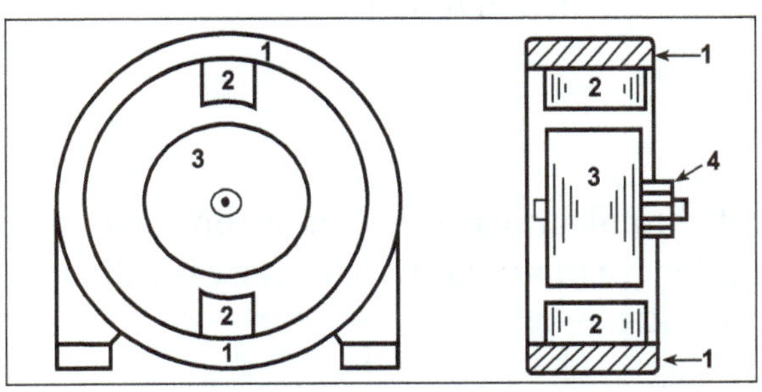

(c) DC Machine and (d) Cross-section of DC Machine.

In the figure:

- Yoke.

- Pole.
- Armature.
- Commutator.

Output Equations and Main Dimensions of DC Machine

Output equation relates the output and main dimensions of the machine. Actually it relates the power developed in the armature and main dimensions.

E - EMF induced or back EMF.

I_a - Armature current.

φ - Average value of flux/pole.

Z - Total number of armature conductors.

N - Speed in rpm.

P - Number of poles.

A - Number of armature paths or circuits.

D - Diameter of the armature.

L - Length of the armature core.

Power developed in the armature in kW $= EI_a \times 10^{-3}$

$$= (\phi ZNP / 60A) \times I_a \times 10^{-3}$$

$$= (P\phi) x \, (I_a Z/A) \times N \times 10^{-3}/60 \qquad ...(1)$$

The term Pφ represents the total flux and is called the magnetic loading. Magnetic loading/unit area of the armature surface is called the specific magnetic loading or average value of the flux density in the air gap B_{av}. That is:

$$B_{av} = P\phi/\pi DL \text{ Wb/m}^2 \text{ or tesla denoted by T.}$$

Therefore, $P\phi = B_{av} \pi DL$ ...(2)

The term $(I_a Z/A)$ represents the total ampere-conductors on the armature and is called the electric loading. Electric loading/unit length of armature periphery is called the specific electric loading q. That is,

$$q = \frac{I_a Z}{A\pi D} \text{ampere} - \text{conductors/m}$$

Therefore, $I_a Z/A = q\pi D$ ...(3)

Substitution of equations (2) and (3) in (1), leads to:

$$kW = B_{av}\pi DL \times q\pi D \times \frac{N \times 10^{-3}}{60}$$

$$= 1.64 \times 10^{-4}\ B_{av}\ qD^2LN$$

$$= C_o D^2 LN$$

Where:

C_o is called the output co-efficient of the DC machine and is equal to $1.64 \times 10^{-4}\ B_{av}q$.

$$D^2L = \frac{kW}{1.64 \times 10^{-4} B_{av} qN} m^3$$

The above equation is called the output equation. The D^2L product represents the size of the machine or volume of iron used. In order that the maximum output is obtained/kg of iron used, D^2L product must be as less as possible. For this, the values of q and B_{av} must be high.

Effect of Higher Value of q

Since, armature current I_a and number of parallel paths A are constants and armature diameter D must be as less as possible or D must be a fixed minimum value, the number of armature conductors increases as $q = I_a Z/A\pi D$ increases.

- As q increases, the number of conductors increases and as a result there will be an increase in conductors/slot, quantity of insulation in the slot, temperature, losses and reduction in heat dissipation and efficiency of the machine.
- As q increases, number of conductors increases and as a result there will be an increase in resistance, I^2R loss and temperature of the machine. Temperature is a limiting factor of any equipment or machine.
- As q increases, number of conductors increases and as a result there will be an increase in armature ampere-turns per pole AT/pole = ($I_a Z/2AP$), flux produced by the armature and effect of armature reaction. In order to overcome the effect of armature reaction there has to be an increase in field MMF. This calls for additional copper and increases the cost and size of the machine.
- As q increases, number of conductors and turns increases leading to increase in reactance voltage proportional to $(\text{turns})^2$. This leads to sparking commutation.

Effect of Higher Value of B_{av}

- As B_{av} increases, the core loss increases and thus efficiency reduces.
- As B_{av} increases, degree of saturation increases, MMF required for the magnetic circuit increases. This calls for additional copper and increases the cost of the machine.

It is clear that there is no advantage gained by selecting higher values of q and B_{av}. If the values selected are less, then D^2L will be large or the size of the machine will unnecessarily be high. Hence optimum value of q and B_{av} must be selected.

In general, q lies between 15000 and 50000 ampere-conductors/m.

Lesser values are used in low capacity, low speed and high voltage machines. In general B_{av} lies between 0.45 and 0.75 T.

Separation of D^2L Product

Knowing the values of kW and N and assuming the values of q and B_{av}, a value for $D^2L = kW/1.64\times10^{-4}\times B_{avq}$ N can be calculated.

Let it be 0.1 m^3.

Since, the above expression has two unknowns namely D and L, another expression relating D and L must be known to find out the values of D and L.

Usually, a value for the ratio armature core length L to pole pitch is assumed to separate D^2L product. The pole pitch τ refers to the circumferential distance corresponding one pole at diameter D. In practice, L/τ lies between 0.55 and 1.1.

Therefore, $L = (0.55 \text{ to } 1.1)\tau$

$$= (0.55 \text{ to } 1.1)\ \pi D/P$$

If $L/\tau = 1.0$ and $P = 4$, then $L = 1.0\times\pi D/P$

$$= 1.0 \times D/4 = 0.785D$$

Therefore, $D^2 \times 0.785D = 0.1$ or $D = 0.5\text{m}$.

Thus, $L = 0.785\times0.5 = 0.395\text{m}$

Specific Electric Loading

It is defined as the ratio of total number of ampere conductors and the armature periphery at the air gap.

- The choice of specific electric loading depends on the following:
  - Temperature rise.
  - Size of machine.
  - Speed of machine.
  - Armature reaction.

- Voltage.
- Commutation.

- A higher value of AC results in a high temperature rise of windings.
- The temperature rise depends on the type of enclosure and cooling techniques employed in the machine.
- If the speed of machine is high, the ventilation of the machine is better and therefore, greater losses can be dissipated. Thus a higher value of AC can be used for machine having high speed.
- In high voltage machines, large space is required for insulation and therefore there is less space for conductors. This means that in high voltage machines, the space left for conductors is less and therefore we should use a small value of AC.
- In large size machines, it is easier to find the space for accommodating conductors. Hence specific electric loading can be increased with increase in linear dimensions.
- With high values of AC, armature reaction will be severe. To counter this, the field MMF is increased and so the cost of the machine goes high.
- High values of AC worsen the commutation condition in machines. From the point of view of commutation, a small value of AC is desirable. The value of AC usually lies between 15000 to 50000 amp.cond/m.

Specific Magnetic Loading

The average flux density over the air gap of a machine is known as specific magnetic loading.

The factors governing the choice of average gap flux density are:

- The maximum flux density in iron parts of the machine.
- The magnetizing current.
- Iron losses.

The value of air gap flux density should be so chosen such that the flux density at the root of the teeth (where the tooth section is minimum) does not exceed 2.2 T. The maximum gap flux density B_g varies from 0.64 to 1.1 T and the corresponding values of B_{av} are 0.45 to 0.75 T.

Choice of Number of Poles

As the armature current increases, cross-sectional area of the conductor and hence the eddy current loss in the conductor increases. In order to reduce the eddy current loss in the conductor, cross-sectional area of the conductor must be made less or the current/path must be restricted.

For a normal design, current/parallel path should not be more than about 200A. However under enhanced cooling conditions, a current/path of more than 200A is also being used. By selecting a suitable number of paths for the machine, current/path can be restricted and the number of poles for the machine can be decided. While selecting the number of poles, the following conditions must also be considered as far as possible.

In order to decide the number of poles to be used, let us consider the different factors affecting the choice of number of poles based on the use of more number of poles.

Frequency

As the number of poles increases, frequency of the induced EMF, f = PN/120 increases, core loss in the armature increases and therefore efficiency of the machine decreases.

Weight of the Iron used for the Yoke

Since the flux carried by the yoke is approximately $\varphi/2$ and the total flux $\phi_T = p\phi$ is a constant for a given machine, flux density in the yoke.

$$B_y = \frac{\phi/2}{\text{cross sectional area of the yoke } A_y} = \frac{\phi_r}{2PA_y} \propto \frac{1}{PA_y}$$

It is clear that A_y is α 1/P as B_y is also almost constant for a given iron. Thus, as the number of poles increases, A_y and hence the weight of iron used for the yoke reduces.

Weight of iron used for the armature core (from the core loss point of view):

Since the flux carried by the armature core is $\varphi/2$, eddy current loss in the armature core α $B_c^2 f^2$.

$$\propto \left[\frac{\phi/2}{A_c}\right]^2 f^2 \propto \left[\frac{\phi_T}{2PA_c}\right]^2 \times \left[\frac{PN}{120}\right]^2$$

$$\propto \frac{1}{A_c^2} \text{ is independent of the number of poles.}$$

On the other hand, since the hysteresis loss in the armature core is given by:

$$\propto B_c^{1.6} f \propto \left(\frac{\phi_T}{2PA_c}\right)^{1.6} \times \frac{PN}{120} \propto \frac{1}{P^{0.6} A_c^{1.6}} \text{, the armature core area } A_c.$$

$\propto \frac{}{0.6/1.6}$ decreases as the number of poles increases for a given hysteresis loss. Thus A_y the weight of iron used for the armature core reduces as the number of poles increases.

Weight of Overhang Copper

For a given active length of the coil, overhang α pole pitch π D/P goes on reducing as the number of poles increases. As the overhang length reduces, the weight of the inactive copper used at the overhang also reduces.

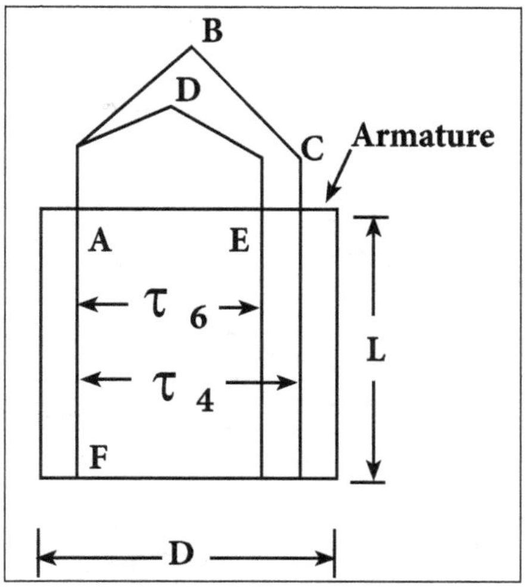

In the figure:

- AF: Active length of the coil.
- ABC or ADE: Overhang or inactive part of the coil.
- τ_4: Pole pitch in case of 4 pole machine.
- τ_6: Pole pitch in case of 6 pole machine.

Armature Reaction

Since the flux produced by the armature $\phi_a = \frac{AT_a/\text{pole}}{\text{reluctance}}$ and armature ampere turns $AT_a/\text{pole} = I_a Z/2AP$ is proportional to 1/P, ϕ_a reduces as the number of poles increases. This in turn reduces the effect of armature reaction.

Overall Diameter

When the number of poles is less, AT_a/pole and hence, the flux produced by the armature is more. This reduces the useful flux in the air gap. In order to maintain a constant value of air gap flux, flux produced by the field or the field ampere-turns must be increased. This calls for more field coil turns and size of the coil defined by the depth of the coil d_f and height of the coil h_f increases.

In order that the temperature rise of the coil is not more, depth of the field coil is generally restricted. Therefore, height of the field coil increases as the size of the field coil or the number of turns of the coil increases.

As the pole height is proportional to the field coil height, height of the pole and hence the overall diameter of the machine increases with the increase in height of the field coil. It is clear that, as the number of poles increases, height of the pole and hence the overall diameter of the machine decreases.

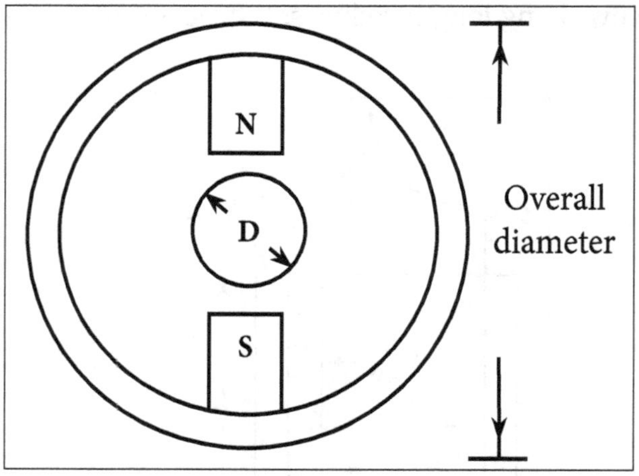

Diameter in case of 2 pole machine.

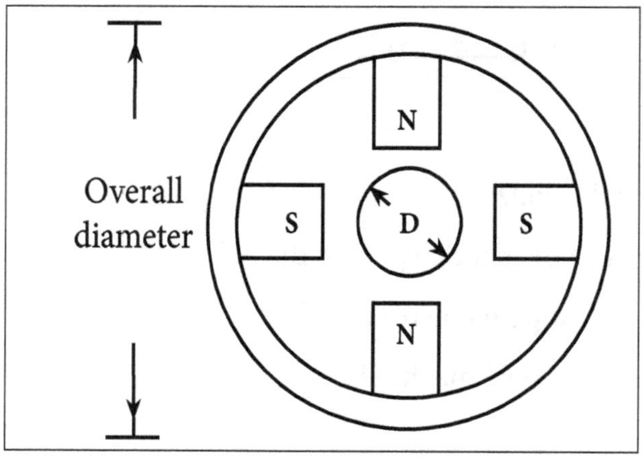

Diameter in case of 4 pole machine.

Length of the Commutator

Since each brush arm collects the current from every two parallel paths, current/brush arm = 2 I_a/A and the cross-sectional area of the brush/arm,

$$A_b = 2I_a/A\delta_b = 2I_a/P\delta_b$$

1/P reduces as the number of poles increases.

As, $A_b = t_b w_b n_b$ and t_b is generally held constant from the commutation point of view, $w_b n_b$ reduces as A_b reduces. Hence the length of the commutator is given by:

$L_c = (w_b n_b + \text{clearances})$ reduces as reduces or the number of poles increases.

Where,

w_b = Width of the brush

t_b = Thickness of the brush

n_b = Number of brushes per spindle

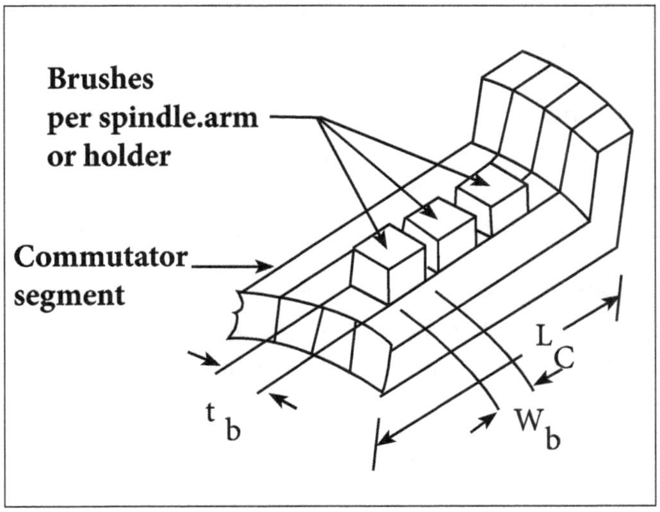

Portion of the commutator.

Flash Over

As the number of poles increases, voltage between the segments increases. Because of the increased value of E_b and carbon dust collected in the space where the mica is undercut, changes of arcing between commutator segment increases. The arc between the segment in turn may bridge the positive and negative brushes leading to a dead short circuit of the armature or flash over.

$$E_b = \frac{\text{voltage between positive and negative brushes}}{\text{number of segaments/pole}}$$

Labour Charges

As the number of pole increases, the cost of labour increases. Since more number of poles are to be assembled, more field coils are to be wound, placed on the pole, insulated, interconnected, etc.

It is clear that, when the number of poles is more, weight of iron used for yoke and armature core, weight of inactive copper, overall diameter, length of commutator and effect of armature reaction reduces. On the other hand, efficiency reduces the chances of flash and cost of machine increases.

Since the advantages outnumber the disadvantages, more number of poles are preferable.

Though more number of poles are preferable, it is not advisable from the cost point of view. In general, the number of poles should be selected so that good operating characteristics are obtained with minimum weight of active material and minimum cost of construction.

Design of Main Dimensions of the DC Machines

In case of DC generator, kW output is considered, in case of DC motor, kW input is considered in finding out the main dimensions D and L from the output equation. In case of DC machine, the speed is specified and the number of poles for the same speed may vary.

Choosing specific magnetic and specific electric loadings for the size of the machine, D^2L product is obtained. The length of the armature core may be taken to be between 0.5 to 1.0 pole pitch. Generally, a smaller diameter and larger core length is used.

When designing a line of DC machine, the armature diameters should be so chosen that a number of ratings of the machine are possible with the same armature diameter. There is no definite rule regarding the minimum and maximum length of the armature core for a given diameter. This will depend on the cost of construction and desired operation characteristics.

To obtain the best proportions of a DC machine, it is necessary to consider the electric and magnetic circuits. It is necessary to choose a suitable number of poles and suitable proportions of them. Regarding electrical circuits, it is necessary to choose a satisfactory winding and commutator.

The current density used in the conductors is between 4.5 to 10 A/mm^2. The class of insulation used for DC machines is B or preferably, F and H (particularly for traction motors).

Air Gap

In case of DC machines, the air gap used is between 1 mm to 6 mm. Normal standard air gaps for medium sized machines are 1.5 mm and 1.75 mm.

Material used for Armature Cores and Poles

For armature, Lohys or special Lohys punchings of 0.5 mm thickness are used. For poles and yokes, CRCA laminated sheets of 1.6 mm thickness are used.

Design of Armature Slot Dimensions

The armature windings can be classified as concentrated and distributed windings.

In case of concentrated windings, all the conductors/pole is housed in one slot. Since the conductor/slot is more, quantity of insulation in the slot is more, heat dissipation is less, temperature rise is more and the efficiency of operation will be less. Also, EMF induced in the armature conductors will not be sinusoidal.

Therefore:

- Design calculations become complicated.
- Core loss increases and efficiency reduces.
- Communication interference may occur.

Hence, no concentrated winding is used in practice for a DC machine armature. In a distributed winding, conductor/pole is distributed in more number of slots.

The distributed winding can be classified as:

- Single layer winding.
- Double layer winding.

In single layer winding, there will be only one coil side in the slot having any number of conductors, odd or even integer depending on the number of turns of the coil. In double layer winding, there will be two or multiples of two coil sides in the slot arranged in two layers. Obviously conductor/slot in a double layer winding must be an even integer.

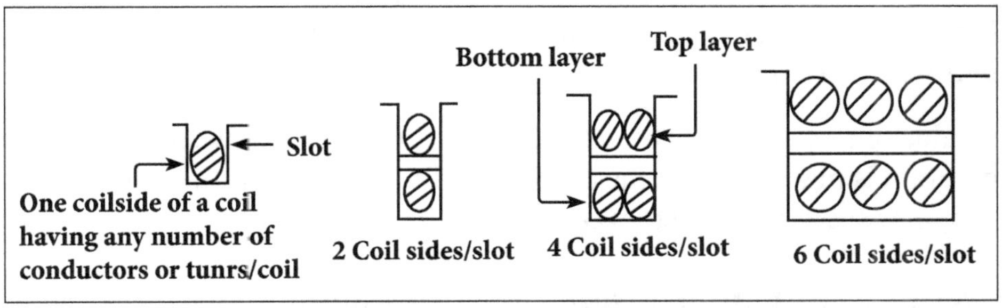

Single layer winding and Double layer winding.

Since for a given number of conductors, poles and slots of a single layer winding calls for less number of coils of more number of turns, reactance voltage proportional to $(\text{turn})^2$ is high. This decreases the quality of commutation or leads to sparking commutation. Hence, a single layer winding is not generally used in DC machines. However, it is much used in alternators and induction motors, where there is no commutation.

Since, a double layer winding calls for more number of coils of less number of turns/coil, reactance voltage proportional to $(\text{turn})^2$ is less and the quality of commutation is good. Hence, double layer windings are much used in DC machines.

Choice between Lap and Wave Winding

Number of Armature Conductors

Since the induced or back EMF, $E = \frac{\phi ZNP}{60A}$, number of armature conductors:

$$Z = \frac{60EA}{\phi NP}$$

Where,

$$\phi = \frac{B_{av}\pi DL}{P}\text{Wb}$$

For a preliminary design, in case of a generator, induced EMF can be taken as the terminal voltage and in case of a motor, back EMF can be taken as the applied voltage.

The number of armature conductors can also be calculated by considering the specific electric loading. In that case, $Z = \frac{qA\pi D}{I}$, the number of conductors must be an even integer.

Cross-sectional Area of the Armature

Since the armature conductors are connected in series in any of the parallel paths of the DC machine, maximum value of the current to be carried by the conductor is I_a/A and the cross-sectional area of the conductor $a = I_a/A\delta mm^2$. In practice, the current density δ lies between 4.5 and 7.0 A/mm².

As the current density increases, the cross-sectional area decreases, resistance of the conductor increases, I^2R loss increases and the temperature of the machine increases. If the cooling facility is not good, then the temperature rise will be high.

Hence, higher value of current density needs to be used in machines where the peripheral velocity or speed of operation is high or where the cooling facility is good. For the calculated area a, round, square or rectangular section conductors can be used.

Choice between Round and Rectangular Conductors

By considering the coil space factor, the type of preferable conductor section can be decided. Coil space factor S_f is defined as the ratio of the area of copper in the coil to the area of the coil.

$$S_f = \frac{\text{Area of copper in the coil } a_f T_f}{\text{overall area of the coil } h_f d_f} < 1.0$$

Where,

a_f = Cross-sectional area of copper or conductor.

T_f = Number of turns or copper sections in the coil.

d_f = Depth of the coil.

H_f = Height of the coil.

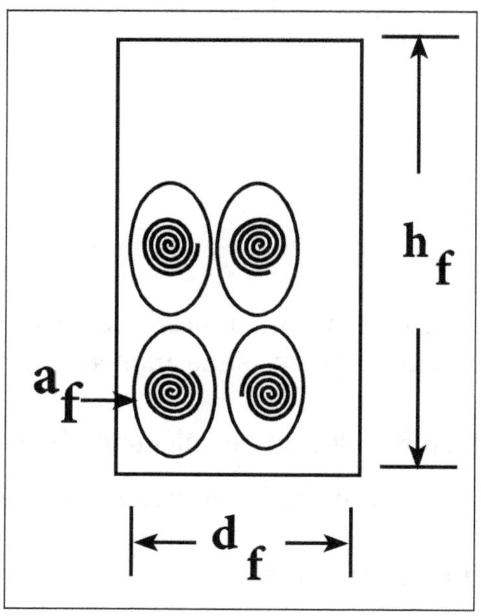

Coil details.

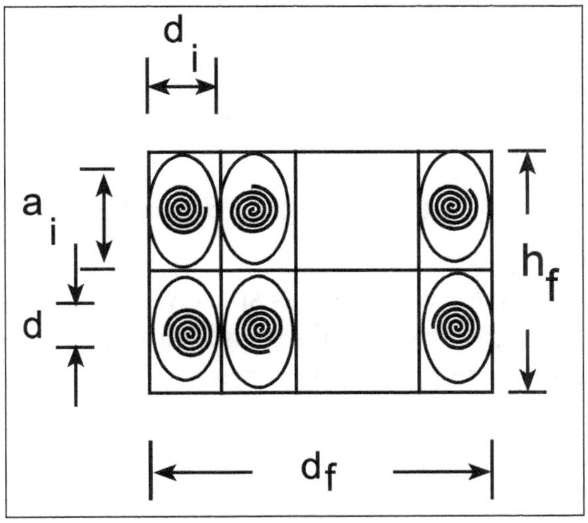

Coil with round conductor.

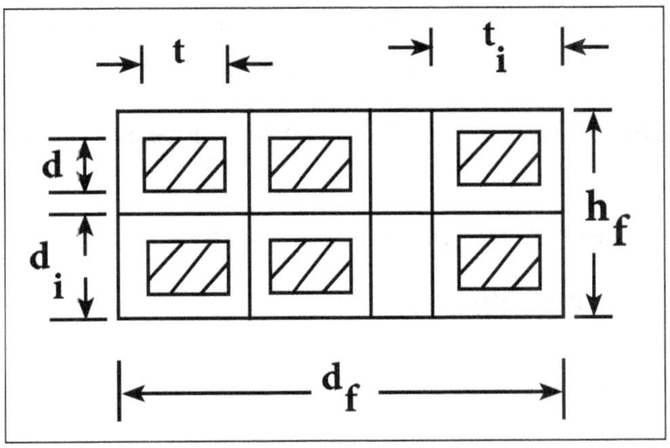

Coil with rectangular conductor.

If d is the diameter of the bare conductor and d_i is the diameter with insulation, then in case of round conductor:

$$S_f = \frac{(\pi d^2/4)\times T_f}{d_i^2 \times T_f} = 0.785(d/d_i)^2$$

S_f is always less than 0.785. In practice, in case of round conductors, S_f lies between 0.3 and 0.65.

If d and t are the depth and width of the rectangular/square conductor without insulation and d_i and t_i are the depth and width with insulation, then in case of a rectangular/square conductor:

$$S_f = S_f = \frac{dt \times T_f}{d_i t_i \times T_f} = \frac{dt}{d_i t_i}$$

In this case, S_f lies between 0.7 and 0.85.

Note: Since the rectangular conductors can be wound closely without wasting any space in between conductors, S_f in case of a rectangular conductor is more than that of a round conductor of same cross-sectional area.

If $a_f = 100$ mm² and $T_f = 48$, the area of coil can be written as:

$h_f d_f = \frac{a_f T_f}{S_{\rightarrow}} = \frac{100 \times 48}{0.6} = 8000\,\text{mm}^2$, in case of round conductors with the assumption that

$S_f = 0.6. = \frac{100 \times 48}{0.6} = 6000\,\text{mm}^2$, in case of rectangular conductors with the assumption that $S_f = 0.8$.

It is clear that, rectangular conductor calls for less space when compared to round conductors of same cross-sectional area. Though a rectangular conductor is preferable, it is costlier than the round conductors. Hence, rectangular conductors are used when the conductor cross-sectional area is more than about 10 mm².

Number of Armature Slots

All DC machines are assumed to have a double layer winding. A double layer winding can be further classified as simplex or multiplex and lap or wave winding. In order to decide the number of slots to be used, the following merits and demerits are considered.

As the number of slot increases, cost of punching the slot increases, number of coils increases and hence, the cost of machine increases.

As the number of slot increases, the slot pitch can be written as:

$$\lambda_s = (\text{Slotwidth } b_s + \text{Tooth width } b_t)$$

$$\lambda_s = \pi D / \text{Number of slots}$$

Here, S decreases and hence the tooth width reduces. This makes the tooth mechanically weak and increases the flux density and core loss in the tooth. Therefore, efficiency of the machine decreases.

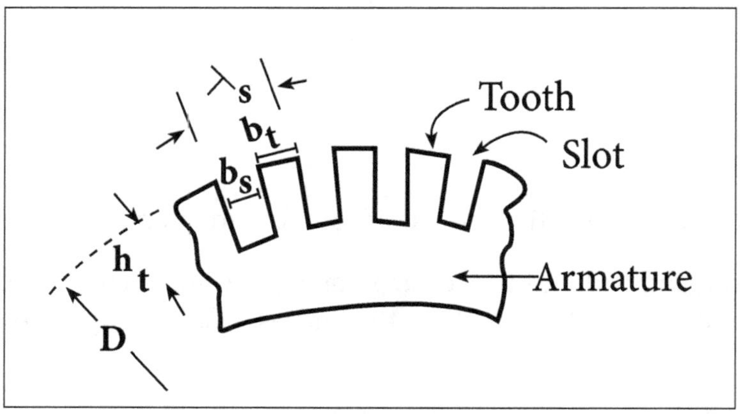

Armature design view.

If the slots are less in number, then the cost of punching and number of coils decrease, slot pitch increases, tooth becomes mechanically strong and efficiency increases, quantity of insulation in the slot increases, heat dissipation reduces, temperature increases and hence the efficiency decreases.

As a preliminary value, the number of slots can be selected by considering the slot pitch. The slot pitch can assume to be between 2.5 and 3.5 cm. This range is applicable only to medium capacity machines and it can be more or less for other capacity machines.

The selection of the number of slots must also be based on the type of winding used, quality of commutation, flux pulsation etc. When the number of slot per pole is a whole number, the number of slots embraced by each pole will be the same for all positions of the armature. However, the number of teeth per pole will not be the same.

This causes a variation in reluctance of the air gap and the flux in the air gap will pulsate. Pulsations of flux in the air gap produce iron losses in the pole shoe and give rise to magnetic noises. On the other hand, when the slots per pole, which is equal to a whole number plus half the reluctance of the flux path per pole pair remains constant for all positions of the armature, there will be no pulsations or oscillations of flux in the air gap.

To avoid pulsations and oscillations of flux in the air gap, the number of slots per pole should be a whole number plus half. When this is not possible, the number of slots per pole arc should an integer.

Armature Core Depth DC

Flux density in the armature core $B_c = \dfrac{\phi/2}{d_c L_i}$ lies between 1.0T and 1.5T. With an assumed value of

$B_c, d_c = \dfrac{\phi/2}{B_c L_i}$ can be calculated.

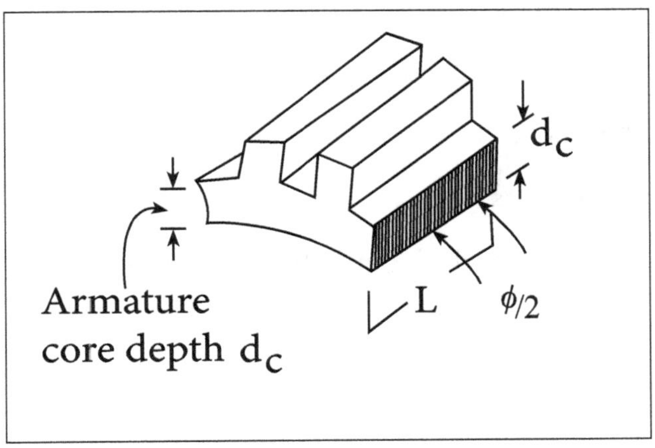

Therefore, the internal diameter of the armature = $(D - 2h_t - 2d_c)$.

Commutator

The commutator is an assembly of commutator segments or bars tapered in section. The segments made of hard drawn copper are insulated from each other by mica or micanite and the usual thickness is about 0.8 mm. The number of commutator segments is equal to the number of active armature coils.

Generally, the diameter of the commutator will be about 60 to 80% of the armature diameter. Lesser values are used for high capacity machines and higher values for low capacity machines.

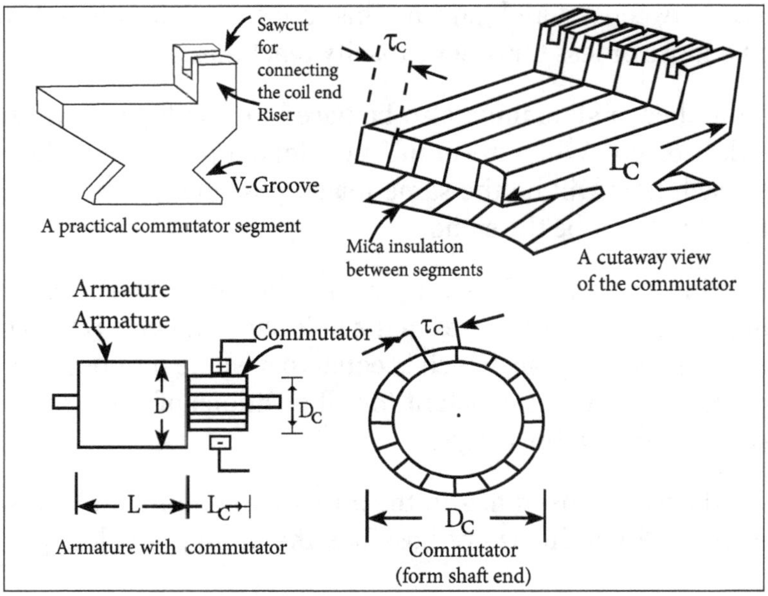

Higher values of commutator peripheral velocity are to be avoided as it leads to lesser commutation time d_t, increased reactance voltage $RV = L(d_i/d_t)$ and sparking commutation. The commutator peripheral velocity $v_c = \pi D_C N/60$ should not be more than about 15 m/s.

The commutator segment pitch τ_C = (Outside width of one segment + Mica insulation between segments) = πD_C/Number of segments should not be less than 4 mm. In practice, the outer surface width of the commutator segment lies between 4 and 20 mm.

The axial length of the commutator depends on the space required:

- By the brushes with brush boxes.
- For the staggering of brushes.
- For the margin between the end of commutator and brush.
- For the margin between the brush and riser and width of riser.

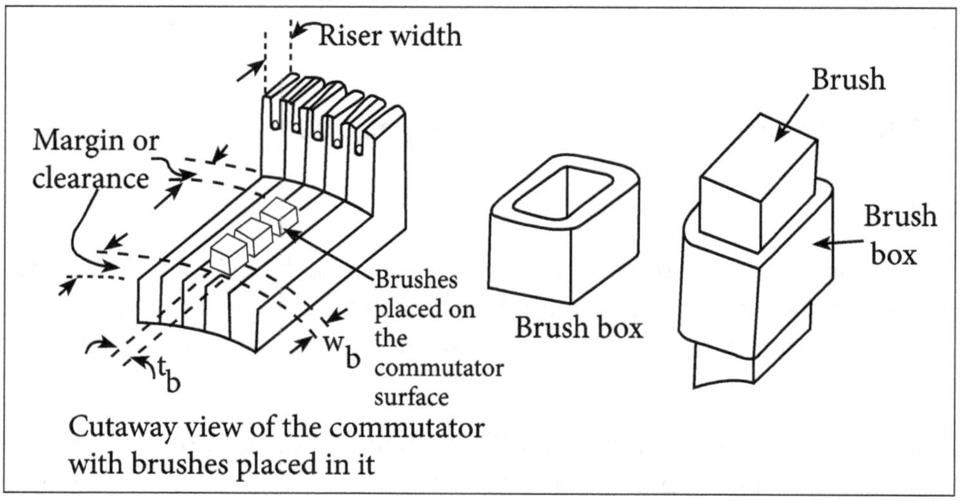

If the length of the commutator leads to a small dissipating surface $\pi D_C L_C$, then the commutator length must be increased, so that the temperature rise of the commutator does not exceed a permissible value, say 55°C.

The temperature rise of the commutator can be calculated by using the following empirical formula:

$$\theta°C = \frac{120 \times \text{watt loss/cm}^2 \text{ of dissipating surface } \pi D_c L_c}{1 + 0.1 v_c}$$

The different losses that are responsible for the temperature rise of the commutator are:

- Brush contact loss.
- Brush frictional loss.

Brush contact loss = Voltage drop/Brush set × I_a

The voltage drop/brush set depends on the brush material such as carbon, graphite, electro graphite or metalized graphite. The voltage drop/brush set can be taken as 2.0 V for carbon brushes.

$$\text{Brush frictional loss} = \text{Frictional torque in } N_m \times \text{Angular velocity}$$

$$= \text{frictional force newton} \times \text{distance in metre} \times \frac{2\pi N}{60}$$

$$9.81 \mu P_b A_{ball} \times \frac{D_c}{2} \times \frac{2\pi N}{60} = 9.81 \mu P_b A_{ball} v_c$$

Where,

μ = Co-efficient of friction and depends on the brush material. It lies between 0.22 and 0.27 for carbon brushes.

P_b = Brush pressure in kg/m². It lies between 1000 and 1500.

A_{ball} = Area of brushes of all the brush arms in m².

$= A_b \times$ Number of brush arms.

$= A_b \times$ Number of poles in case of lap winding.

$= A_b \times 2$ or P in case of wave winding.

A_b = Cross-sectional area of the brush/brush arm.

Brush Details

Since the brushes of each brush arm collect the current from two parallel paths, current collected by each brush arm is $2I_a/A$ and the cross-sectional area of the brush or brush arm or holder or spindle $A_b=2I_a/A\delta_b$. The current density depends on the brush material and δ_b can be assumed between 5.5 and 6.5 A/cm² for carbon.

In order to ensure a continuous supply of power, a number of sub-divided brushes are used instead of one single brush. Also the cost of replacement of damaged or worn out brushes is cheaper.

Thus if:

- t_b is the thickness of the brush.
- w_b is the width of the brush.
- n_b is the number of sub-divided brushes.

Then, $A_b = t_b w_b n_b$

As the number of adjacent coils of the same or different slots that are simultaneously undergoing commutation increases, the brush width and time of commutation also increases at the same rate and therefore, the reactance voltage becomes independent of the brush width.

With only one coil undergoing commutation and width of the brush equal to one segment width, the reactance voltage and hence the sparking increases as the slot width decreases. Hence, the brush width is made to cover more than one segment. If the brush is too wide, then those coils which are away from the commutating pole zone or the coils that does not come under the influence of inter-pole flux and undergo commutation leads to sparking commutation.

Hence, brush width greater than the commutating zone width is not advisable under any circumstances. Since the commutating pole zone lies between 9 and 15% of the pole pitch, 15% of the commutator circumference can be considered as the maximum width of the brush.

It has been found that, the brush width should not be more than 5 segments in machines less than 50 kW and 4 segments in machines more than 50 kW.

The number of brushes/spindle can be found out by assuming a standard brush width or a maximum current/sub-divided brush.

Standard brush width can be 1.6, 2.2 or 3.2 cm. Current/sub-divided brush should not be more than 70 A.

Thus, with the assumed brush width, $n_b = A_b / t_b w_b$. With the assumed current/sub-divided brush, $n_b = 2I_a / A*70$ and $w_b = A_b / t_b n_b$.

Staggering of Brushes

During the flow of current from the commutator segment to the brush, copper is eaten away leading to formation of ridges between the sub-divided brushes of the same brush arm. Since it is not possible to avoid eating away of copper by the arc, eating away of copper must be made to take place over the entire axial length of the commutator to ensure uniform commutator surface. This is achieved by displacing all the positive brushes in one direction and all the negative brushes in the other direction or by staggering of brushes in pairs as shown in the figure:

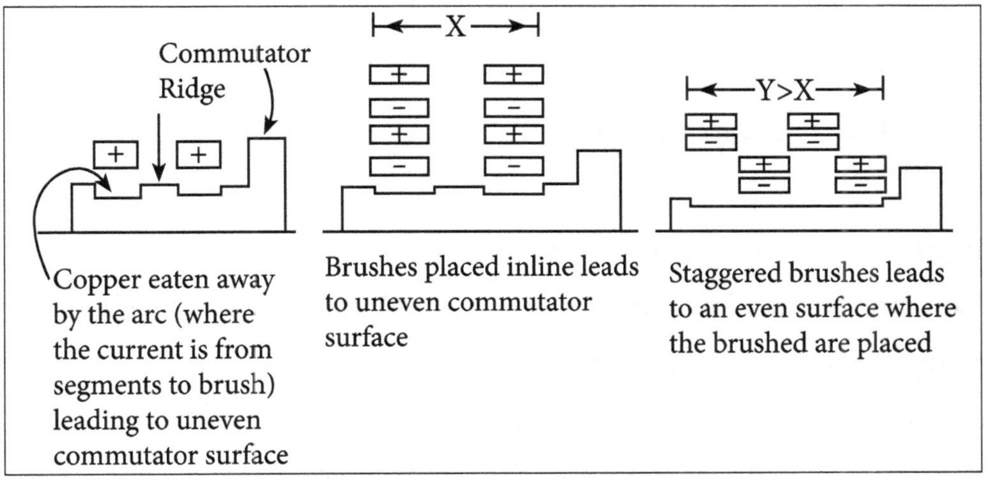

Brush materials and their properties.

Material	Peripheral velocity m/s	Current density in A/cm²	Voltage drop per brush set in volts	Co-efficient of friction
Normal carbon	5 to 15	5.5 to 6.5	2.0	0.22 to 0.27
Soft graphite	10 to 25	9.0 to 9.5	1.6	0.12
Metalized graphite (Copper-carbon mixture)	5 to 15	15 to 16	0.24 to 0.35	0.16
Electro graphite (Graphitized by heating)	5 to 15	8.5 to 9.0	1.7 to 1.8	0.22

Step by step design procedure of commutator and brushes:

- Diameter of the commutator, DC = 0.6D to 0.8D. It must be in such a way that the peripheral velocity of the commutator $v_C = \pi D_C N/60$ is not more than 15 m/s.
- The commutator segment pitch $\tau_C = \pi D_C$/Number of segments should not be less than 4 mm from the mechanical strength point of view.
- The number of commutator segments is equal to number of active armature coils.
- If there are n_b brushes/brush arm or spindle or holder, placed one beside the other on the commutator surface, then the length of the commutator L_C = (Width of the brush w_b + Brush box thickness 0.5 cm) Number of brushes/spindle + End clearance 2 to 4 cm + Clearance for risers 2 to 4 cm + Clearance for staggering of brushes 2 to 4 cm.
- Cross-sectional area of the brush/spindle or arm or holder, $A_b = 2I_a / A\delta_b$. The current density in the brushes δ_b lies between 5.5 and 6.5 A/cm² for carbon brushes.
- Maximum thickness of the brush, $t_b = 4\,\tau_c$ for machines greater than 50 kW = $5\,\tau_c$ for machines lesser than 50 kW.

- With the standard brush width W_b, the number of brushes/spindle is given by:

  $n_b = A_b / t_b W_b$

Total commutator losses = Brush contact loss + Brush frictional loss

$= \text{Voltage drop/Brush set} \times I_a + 9.81\ \mu P_b A_{ball} C_v$

Where,

Voltage drop/set = 2.0 V for carbon brushes.

μ = Co-efficient of friction. It lies between 0.22 to 0.27 for carbon brushes.

P_b = Brush pressure. It lies between 1000 and 1500 kg/m².

Temperature rise of the commutator is given by:

$$\theta°C = \text{Cooling co-efficient} \times \text{Watt loss / dissipating surface}$$

$$= \frac{120}{1+0.1v_c} \times \text{watt loss/cm}^2 \text{ of dissipating surface } \pi D_c L_c$$

Temperature rise should be less than about 55°C.

Problems

1. Let us determine the number of poles, armature diameter and core length for the preliminary design of a 500 kW, 400 V, 600 rpm, DC shunt generator assuming an average flux density in the air gap of 0.7 T and specific electric loading of 38400 ampere-conductors per meter. Assume core length/pole arc = 1.1.

Solution:

Given:

Average flux density in the air gap = 0.7 T

Specific electric loading = 38400 ampere-conductors per meter

Formula to be used:

$$I_a = I_L + I_{sh} \frac{\text{output in watts}}{\text{terminal voltage}}$$

$$f = \frac{PN}{120}$$

$$D^2L = \frac{kW}{1.64 \times 10^{-4} B_{av} q N}$$

$$\tau = \frac{\pi D N}{P}$$

$$v = \frac{\pi DN}{60}$$

Armature current:

$$I_a = I_L + I_{sh} \frac{\text{output in watts}}{\text{terminal voltage}} = \frac{500 \times 10^3}{400} = 400A$$

In order that the current/path is not more than 200A, a lap winding is to be used. The number of parallel paths required is given by:

$$A = \frac{1250}{200} = 6.25$$

Since the number of parallel paths cannot be a fraction as well an odd integer, A can be 6 or 8. Let it be 6. (Note: A current/path of 200A need not strictly be adhered).

Check: Frequency of the induced EMF, $f = \frac{PN}{120} = \frac{6 \times 60}{120} = 30\text{Hz}$ and can be considered as the frequency generally lies between 25 and 50 Hz.

$$D^2L = \frac{kW}{1.64 \times 10^{-4} B_{av} qN}$$

$$= \frac{500}{1.64 \times 10^{-4} \times 0.7 \times 38400 \times 600}$$

$$\approx 0.189\text{m}^3$$

Since the capacity is considerable, power developed in the armature can be taken as power output for a preliminary design.

$$L = 1.1 \text{ pole arc} = 1.1^{\psi\tau}$$

Since, Ψ lies between 0.6 and 0.7, let it be 0.7.

Since, $\tau = \frac{\pi D}{P}, L = 1.1 \times 0.7 \times \frac{\pi D}{6} = 0.4D$

$$D^2 \text{x } 0.4D = 0.189$$

$$D^3 = \frac{0.189}{0.4} = 0.473\text{m}^3$$

$$D = 0.78\text{m and } L = 0.4 \times 0.78 \approx 0.31 \text{ m}$$

Check: Peripheral velocity,

$$v = \frac{\pi DN}{60} = \frac{\pi \times 0.78 \times 600}{60} = 24.2\,m/s$$ and is within the limits, (i.e.) 30 m/s.

2. Let us determine the diameter and length of the armature core for a 55 kW, 110 V, 1000 rpm and 4 pole DC shunt generator. We will assume specific magnetic loading 0.5T, Specific electric loading 13000 ampere-turns, pole arc 70% of pole pitch and length of core about 1.1 times the pole arc, 10 A for field current and a voltage drop of 4 V for the armature circuit. Let us also determine the number of armature conductors and slots.

Solution:

Given:

Specific magnetic loading - 0.5T

Specific electric loading - 13000 ampere-turns

Pole arc - 70% of pole pitch

Length of core - 1.1 times the pole arc

Field current - 10 A

Voltage drop - 4 V

Formula to be used:

Power developed in the armature in kW $= EI_a \times 10^{-3}$

$$E = V + \text{Voltage drop in the armature circuit}$$

$$I_a = I_L + I_{sh}$$

$$D^2L = \frac{kW}{1.64 \times 10^{-4} B_{av} q N}$$

$$Z = \frac{60EA}{\varphi NP}$$

$$\varphi = \frac{B_{av} \pi DL}{P}$$

$$\lambda_s = \frac{\pi D}{S}$$

Power developed in the armature in kW $= EI_a \times 10^{-3}$

$$E = V + \text{Voltage drop in the armature circuit} = 110 + 4 = 114V$$

$$I_a = I_L + I_{sh} = \frac{55 \times 10^3}{110} + 10 = 510A$$

$$kW = 114x\ 510\ x10^{-3} = 58.14$$

$$q = 13000 \times\ 2 = 26000$$

As q must be substituted in ampere conductors and one turn corresponds 2 conductors:

$$D^2L = \frac{58.14}{1.64 \times 10^{-4} \times 0.5 \times 26000 \times 1000}$$

$$\approx 0.027\ m3$$

Since L = 1.1 times pole arc, $L = \Psi\tau = 1.1 \times 0.7 \times \pi D/4 = 0.4D$

$D^3 = \dfrac{0.027}{0.6} m^3$ and therefore $D = 0.36m$ and $L = 0.6\ \times\ 0.36 \approx 0.22m$

Number of armature conductors is given by:

$$Z = \frac{60EA}{\varphi NP}$$

Flux per pole is given by:

$$\varphi = \frac{B_{av}\pi DL}{P} = \frac{0.5 \times \pi \times 0.36 \times 0.22}{4} = 0.03Wb$$

If a lap winding is assumed, A = P and therefore:

$$Z = \frac{60 \times 114 \times 4}{0.03 \times 1000 \times 4} \approx 228$$

Since the minimum number of slots per pole is 9 from commutation point of view, minimum number of slots for the machine is $= 9\ \times 4 = 36$ ...(1)

Since slot pitch $\lambda_s = \dfrac{\pi D}{S}$, lies between 2.5 and 3.5 cm, the number of slots lies between $\dfrac{\pi \times 36}{3.5} \approx 32$

and $\dfrac{\pi \times 36}{2.5} \approx 45$ ...(2)

From equation (1) and (2), the number of slots lies between 32 and 45.

Since a lap winding is assumed and for a lap winding the number of slots may be multiple of number of poles, (i.e.) 6 or multiple of number of pair poles, (i.e.) 3, let the number of slots S = 40.

Since preliminary number of slots and conductors are known, actual number of conductors per slot and number of conductors can be calculated as shown below.

Conductors per slot = 228/40 = 5.7 and is not possible as conductors per slot must always be an even integer for a double layer winding. Since only double layer winding is for DC machines, let the number of conductors per slot be 6. Therefore, $Z_{revised}$ = 40 x 6 = 240.

3. Let us calculate the armature diameter and core length for a 7.5 kW, 4 pole, 1000 rpm and 220 V shunt motor. Assume: full load efficiency = 0.83, field current is 2.5% of rated current. The maximum efficiency occurs at full load.

Solution:

Given:

Full load efficiency - 0.83

Field current - 2.5% of rated current

Formula to be used:

$$\text{Electrical input to the motor} = \frac{\text{output}}{\eta}$$

$$\text{Losses at full load} = \left(\frac{1-\eta}{\eta}\right)\text{output}$$

$$\text{Field copper loss} = (I_{sh})^2 R_{sh}$$

$$D^2L = \frac{kW}{1.64 \times 10^{-4} B_{av} qN}$$

$$L = \tau = \frac{\pi D}{4}$$

Power developed in the armature = Electrical input to the motor - Field and armature copper losses = Electrical output of the motor + Iron friction and windage losses

$$\text{Electrical input to the motor} = \frac{\text{output}}{\eta} = \frac{7.5 \times 10^3}{0.83} \approx 9036\text{W}$$

$$\text{Losses at full load} = \left(\frac{1-\eta}{\eta}\right)\text{output}\left(\frac{1-0.83}{0.83}\right) \times 7.5 \times 10^3 = 1536\text{W}$$

Since efficiency is maximum at full load and at maximum efficiency, the variable loss is equal to constant loss:

$$\text{variables loss} = \text{armature copper lose} = \frac{1536}{2} \approx 768\text{W}$$

Field copper loss = $(I_{sh})^2 R_{sh}$ or VI_{sh}

$$I_{sh} = 0.025 \text{ full load current} = 0.025 \times \frac{7.5 \times 10^3}{0.83 \times 220} = 1.03\text{A}$$

Field copper loss = $220 \times 1.03 = 226.6\text{W}$

Power developed in the armature $= 9036 - 768 - 226.6 = 8041.4$ W

Iron, friction and windage losses = Constant losses - Field copper loss = 768 - 226.6 = 541.4W

Power developed in the armature = $7500 + 541.4 = 8041.4\text{W}$

Since specific magnetic loading lies between 0.45 and 0.75 T, let it be 0.6T. Since specific electric loading lies between 15000 to 50000 ampere-conductors, let it be 30000.

$$D^2L = \frac{8.041}{1.64 \times 10^{-4} \times 0.6 \times 30000 \times 1000} = 0.0024\,\text{m}^3$$

Since L/τ lies between 0.55 and 1.1, let it be 1.0.

Therefore:

$$L = \tau = \frac{\pi D}{4} = 0.785D$$

4. Let us determine the main dimensions, number of poles, number of conductors per slot and air gap length subjecting the results to design checks for a 600 kW, 500 V 900 rpm DC generator. Assume: Average flux density 0.6T, ampere-conductors per meter 35000. The ratio of pole arc to pole pitch is 0.67 and efficiency is 91%.

Peripheral velocity should not exceed 40m/s.

Armature MMF per pole should be below 7500A.

Current per brush arm should not exceed 400A.

Frequency of the flux reversal should not exceed 50Hz.

5. The MMF required for air gap is 50% of armature MMF and gap contraction factor is 1.15.

Solution:

Given:

Average flux density - 0.6T

Ampere-conductors per meter – 35000

The ratio of pole arc to pole pitch - 0.67

Efficiency - 91%.

Peripheral velocity should not exceed 40m/s.

Armature MMF per pole should be below 7500A.

Current per brush arm should not exceed 400A.

Frequency of the flux reversal should not exceed 50Hz.

MMF required for air gap - 50% of armature MMF

Gap contraction factor - 1.15

Formula to be used:

$$f = \frac{PN}{120}$$

$$D^2L = \frac{kW}{1.64 \times 10^{-4} B_{av} qN}$$

$$kW = \frac{Output}{\eta}$$

$$v = \frac{\pi DN}{60}$$

$$\lambda_s = \frac{\pi D}{S}$$

$$Z = \frac{E60A}{\varphi NP}$$

$$\varphi = \frac{B_{av} \pi DL}{P}$$

$$AT_a/pole = \frac{I_a Z}{2AP}$$

$$B_g = \frac{B_{av}}{\psi}$$

Armature current:

$$I_a \approx I_L = \frac{600 \times 10^3}{300} = 1200A$$

In order that current per path is not greater than 200 A or current per brush arm is not greater than 400 A, a lap winding is to be used having parallel paths:

$$A = \frac{1200}{200} = 6$$

Since for a simplex lap winding, $A = P,\ P = 6$

Check: Frequency $f = \frac{PN}{120} = \frac{6 \times 900}{120} = 45 < 50\text{Hz}$ and the condition are satisfied.

$$D^2L = \frac{kW}{1.64 \times 10^{-4} B_{av} qN}$$

Power developed:

$$kW = \frac{\text{Output}}{\eta} = \frac{600}{0.91} = 659.4$$

$$D^2L = \frac{659.4}{1.64 \times 10^{-4} \times 0.6 \times 3500 \times 900} = 0.213\,m^3$$

Since the peripheral velocity should not be greater than 40 m/s, let it be 36 m/s.

Since:

$$v = \frac{\pi DN}{60}, D = \frac{60v}{\pi N} = \frac{60 \times 36}{\pi \times 900} = 0.76m$$

$$L = \frac{0.213}{0.76^2} = 0.28m$$

Since the minimum number of slots per pole is 9 from commutation point of view, minimum number of slots for the machine is $= 9 \times 6 = 54$ ...(1)

Since slot pitch, $\lambda_s = \frac{\pi D}{S}$, lies between 2.5 and 3.5 cm, the number of slots lies between $\frac{\pi \times 75}{3.5} \approx 68$

and $\frac{\pi \times 76}{2.5} \approx 96$ ...(2)

From equation (1) and (2) the number of slots lies between 68 and 96.

Since for a lap winding the number of slots may be multiple of number of poles, (i.e.) 6 or multiple of number of pair poles, (i.e.) 3, let the number of slots S = 72.

Number of armature conductors:

$$Z = \frac{E60A}{\varphi NP}$$

$$\varphi = \frac{B_{av}\pi DL}{P} = \frac{0.6 \times \pi \times 0.76 \times 0.28}{6} = 0.067\,Wb$$

$$Z = \frac{60 \times 500 \times 6}{0.067 \times 900 \times 6} \approx 498$$

Conductors per slot $= \frac{498}{72} = 6.9$ and is not possible. Let it be 6.

$$Z_{revised} = 72 \times 6 = 432$$

Armature MMF per pole $AT_a/\text{pole} = \frac{I_a Z}{2AP} = \frac{1200 \times 432}{2 \times 6 \times 6} = 7200A < 7500A$ and the condition is satisfied.

Air gap MMF is given by:

$$AT_g = 800000 l_g K_g B_g = 0.5\, AT_a/\text{pole}$$

Carter's gap expansion or contraction co-efficient $K_g = 1.15$

Maximum gap density is given by:

$$B_g = \frac{B_{av}}{\psi} = \frac{0.6}{0.67} = 0.895T$$

Air gap length is given by:

$$I_g = \frac{0.50 \times 7200}{800000 \times 1.15 \times 0.895} = 0.0043m$$

6. A 500 kW, 500 V, 375 rpm, 8 pole DC generator has an armature diameter of 110 cm and the number of armature conductors is 896. Let us calculate the diameter of the commutator, length of the commutator, number of brushes per spindle, commutator losses and temperature rise of the commutator. Assume single turn coil.

Solution:

Given:

Armature diameter = 110 cm

Number of armature conductors = 896

Formula to be used:

$$\text{Armature current } I_a = \frac{kW \times 10^3}{V}$$

$$Ab = 2I_a / A\delta_b$$

Commutator segment pitch $\tau_c = \pi D_c$/Number of segments or coils

Number of coils = Z/2 x Number of turns per coil:

$$n_b = \frac{A_b}{t_b W_b}$$

Brush contact loss = Voltage drop/Brush set x I_a

Brush frictional loss = 9.81 $\mu P_b A_{ball} v_c$

$$\theta = \frac{120 \times \text{watt loss/cm}^2 \text{of the commutator dissipating surface } \pi D_c L_c}{1 + 0.1 v_c}$$

Diameter of the commutator $DC = 0.6D$ to $0.8D = 0.7 \times 110 = 77$ cm

Length of the commutator L_c = (Width of the brush W_b + Brush box thickness 0.5 cm) Number of brushes/spindle n_b + End clearance 2 to 4 cm + Clearance for risers 2 to 4 cm + Clearance for staggering of brushes 2 to 4 cm.

$$\text{Armature current } I_a = \frac{kW \times 10^3}{V} = \frac{500 \times 10^3}{500} = 1000\text{A}$$

An armature current of 1000 A obviously calls for a lap winding.

Cross-sectional area of the brush per spindle or brush arm or holder $A_b = 2I_a / A\delta_b$, since the current density lies between 5.5 and 6.5 A/cm² for carbon brushes, let it be 6 A/mm².

$$A_b = \frac{2 \times 1000}{8 \times 6} = 41.66\text{cm}^2$$

Maximum thickness of the brush = $4\tau_c$

Commutator segment pitch $\tau_c = \pi D_c$/Number of segments or coils

Number of coils = Z/2 x Number of turns per coil $= 896/2 \times 1 = 448$

Therefore:

$$\tau = \frac{\pi \times 77}{448} = 0.54\text{cm}$$

Maximum thickness of the brush $= 4 \times 0.54 = 2.16$ cm

Let the thickness of the brush t_b = 2.0 cm

If a brush width of 2.2 cm is assumed, then W_b = 2.2 cm

Thus, number of brushes/spindle is given by:

$$n_b = \frac{A_b}{t_b W_b} = \frac{41.66}{2 \times 2.2} = 9.46 \text{ and is not possible.}$$

Let the number of brushes/spindle be = 10

Therefore, $L_c = (2.2 + 0.5)10 + 2 + 2 + 2 = 33$ cm

Brush contact loss = Voltage drop / Brush set $\times I_a = 2 \times 1000 = 2000$ W

Brush frictional loss = $9.81\ \mu P_b A_{ball} v_c$

Let the brush pressure P_b = 1215 kg/m² as it lies between 1000 to 1500 kg/m²

A_{ball} = Area of brushes of all the brush arms

= $t_b w_b n_b$ x Number of brush arms or number of poles

$= 2 \times 2.2 \times 10 \times 8 \times 10^{-4} = 0.0352\ m^2$

Brush frictional loss $= 9.81 \times 0.25 \times 1215 \times 0.352 \times 15.1 = 1583.8$ W

Therefore commutator losses (total) $= 2000 + 1583.8 = 3583.8$ W

Temperature rise in degree centigrade is given by,

$$\theta = \frac{120 \times \text{watt loss/cm}^2 \text{of the commutator dissipating surface } \pi D_c L_c}{1 + 0.1 v_c}$$

$$= \frac{120 \times 3583.8 / \pi \times 77 \times 33}{1 + 0.1 \times 15.1} = 21.46$$

7. A 20 Hp, 4 pole, 250 V, 1000 rpm wave wound DC machine has the following data. Diameter of the armature = 25 cm, number of slots = 41, number of coil sides/slot = 4, turns/coil = 2. Let us calculate the number of segments, outside width of one segment and mica, brush thickness, length of the commutator and brush contact loss.

Solution:

Given:

Diameter of the armature = 25 cm

Number of slots = 41

Number of coil sides/slot = 4

Turns/coil = 2

Formula to be used:

$$\text{Number of coils} = \frac{\text{total number of conductors}}{2 \times \text{number of turns/coil}}$$

Number of coils = Number of slots x Number of coil sides/layer

$$Y_c = \frac{C+1}{p}$$

$$\text{Armature current } I_a = \frac{Hp \times 746/\eta}{V}$$

$$\text{cross-sectional area of the brush/ spindle } A_b = \frac{21_a}{A\delta_b}$$

$$n_b = \frac{A_b}{t_b W_b}$$

Brush contact loss = Voltage drop/brush set x I_a

Number of segments = Number of active coils for a wave winding

- Number of coil sides (total) = 41 x 4 = 164.
    - Number of coils = 164/2 = 82 as each coil will have 2 coil sides.
- Since the coil is of 2 turns, each coil side will have 2 conductors and therefore, the number of conductors per slot = 4 x 2 = 8.

Total number of conductors = 41 x 8 = 328

$$\text{Number of coils} = \frac{\text{total numbers of conductors}}{2 \times \text{number of turns/coil}} = \frac{328}{2 \times 2} = 82$$

i) Number of coils = Number of slots x Number of coil sides/layer = 41 x 2 = 82

For a wave winding, $Y_c = \frac{C \pm 1}{p}$ must be an integer. With the number of coils calculated, $Y_c = \frac{82 \pm 1}{2}$ is a fraction. Hence, a wave winding is not possible. However, a wave winding can be made possible by considering one of the coils as dummy. Therefore, number of active coils = 81 and number of commutator segments = 81.

Outside width of one segment and mica = Commutator segment pitch

$$= \frac{\pi D_c}{\text{number of segments}} = \frac{\pi \times 0.7 \times 25}{81} \text{ with the assumption that DC} = 0.7$$

$$D = 0.68 \text{ cm}$$

Maximum thickness of the brush = 5 times the commutator segment pitch = 5 x 0.68 = 3.4 cm.

Let the thickness of the brush, t_b = 2.5 cm

$$\text{Armature current } I_a = \frac{Hp \times 746/\eta}{V} = \frac{20 \times 746/0.9}{250} = 66.3A$$

$$\text{cross - sectional area of the brush/spindle } A_b = \frac{2I_a}{A\delta_b} = \frac{2 \times 66.3}{2 \times 6} = 11.65 \text{cm}^2$$

Let the standard brush width, $W_b = 1.6\text{cm}$

$$\text{Number of brushes/spindle } n_b = \frac{A_b}{t_b W_b} = \frac{11.65}{2.5 \times 1.6} = 2.76 \text{ and is not possible}$$

$$= 3 \text{ (say)}$$

Length of the commutator, $L_C = (1.6 + 0.5)3 + 2 + 2 + 2 = 12.3$ cm

Brush contact loss = Voltage drop / brush set $\times I_a = 2.0 \times 66.3 = 132.6$ W

1.2 3-phase Synchronous Machine with and without Damper Bars and 3-phase Induction Machine

Damper Winding

Synchronous motors are not self-starting machines. Synchronous motors are made self-starting by providing a special winding in the rotor poles known as damper winding or squirrel cage windings.

The damper winding consists of short circuited copper bars embedded in the face of the rotor poles. When an AC supply is provided to the stator of a 3-phase synchronous motor, stator winding produces rotating magnetic field.

Due to the damper winding present in the rotor winding of the synchronous motor, machine starts as induction motor. Induction machine works on the principle of induction. The Damper windings in synchronous motor will carry out the same task of induction motor rotor windings. Therefore due to damper windings synchronous motor starts as induction motor and continue to accelerate.

The exciter for synchronous motor moves along with the rotor. When the motor attains about

95% of the synchronous speed, the rotor windings is connected to exciter terminals and the rotor is magnetically locked by the rotating magnetic field of stator and it runs as a synchronous motor.

The damper winding is mostly used in the synchronous machine- in stator of synchronous generator and rotor of synchronous motor. It provides some specific advantage to synchronous machines like smooth starting and compensating the transient effect or unbalanced condition. The rotor of induction machine is mostly of squirrel cage type which is short circuited thus provides high current in startup.

Damper winding in the synchronous machine is placed in the coil slots and short circuited. Thus the Damper winding provides the same effect as the squirrel cage winding rotor does in case of induction motor.

Position of Damper Winding

Damper winding is placed in the slots at the surface of pole shoes in salient pole machines and in non-salient pole machine; it is placed in the same slots with magnetizing windings.

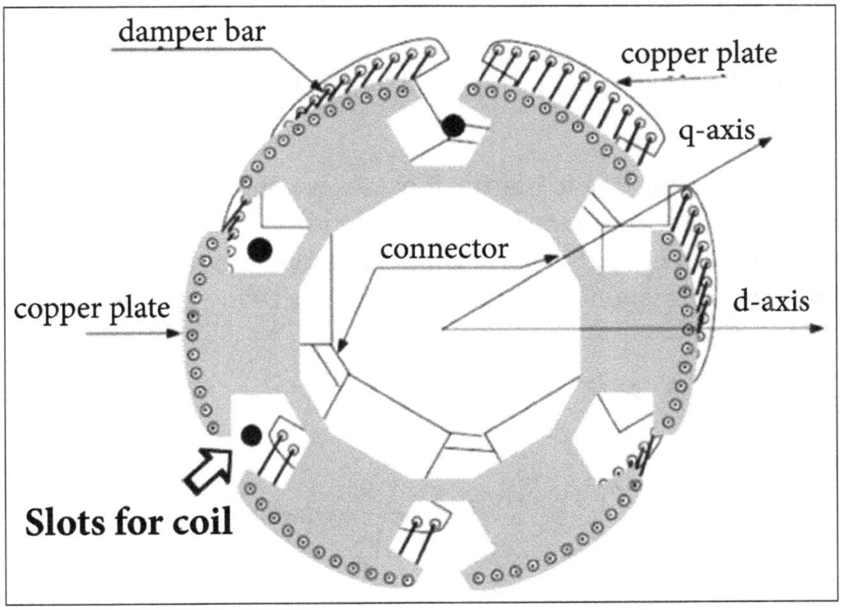

Damper winding in salient pole machine.

Function of Damper Winding in Synchronous Machine

Self-starting of Synchronous Machine

The core concept is that damper winding being short circuited carries high current, creates magnetic flux like as induction machine. So, there is pulling in or better torque in starting. This is done in the moment of starting when synchronous machine is not yet reached in synchronism.

Because the synchronous machine before synchronism has the characteristics of an induction machine as there is relative motion between rotor and stator. When the machine reached synchronism, there will be no current in damper winding as there is no relative motion between rotor and stator.

Compensation of Transients

Compensation of transients by damping the rotational fluctuation (hunting) caused by unbalanced loading, torque etc. Whenever the synchronous machine is pulled out from synchronism as there will be relative motion between the rotor and stator, the damper winding will provide flux that will compensate the transient effect and machine will return back to synchronism.

Turbo Generator usually does not have damper windings, as their solid field poles functions as damper. But they might use damper windings for synchronization purpose.

Damper windings are copper bar slotted in pole shoes of electrical machine. The dampers are short circuited at each ends by heavy copper rings.

Hunting

When synchronous motor is on no load, the stator and rotor pole axes almost coincide with each other. When motor is loaded, the rotor axis falls back with respect to stator.

The angle by which rotor retards are called load angle or angle of retardation δ. If the load connected to the motor is suddenly changed by a large amount, then the rotor tries to retard to take its new equilibrium position. But due to inertia of the rotor, it cannot achieve its final position instantaneously.

While achieving its new position due to inertia, it passes beyond its final position corresponding to the new load. This will produce more torque than demanded. This will try to reduce the load angle and rotor swings in other direction. So, there is periodic swinging of the rotor on each sides of the new equilibrium position corresponding to the load. Such a swing is shown in the figure:

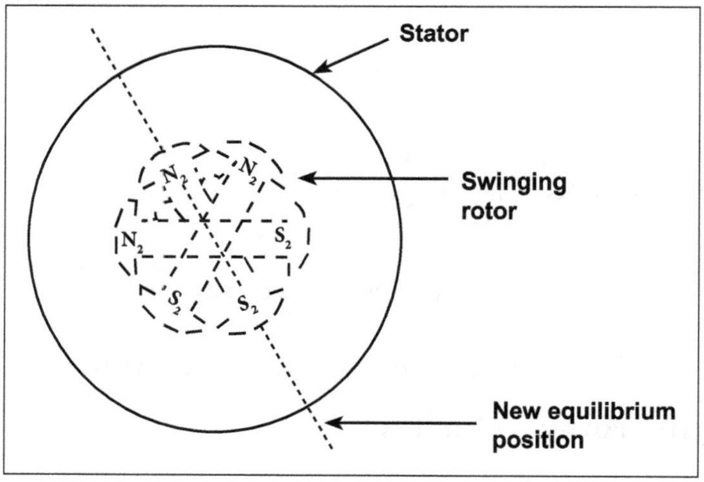

Hunting in synchronous motor.

Such oscillations of the rotor about its new equilibrium position due to sudden application or removal of load is known as swinging or hunting in synchronous motor.

Due to such hunting, the load angle changes its value about its final value δ and the current drawn by the motor also changes. Hence during hunting there are changes in the current drawn by the motor which can cause problem to the other appliances connected to the same line.

The changes in armature current due to hunting are shown in the figure:

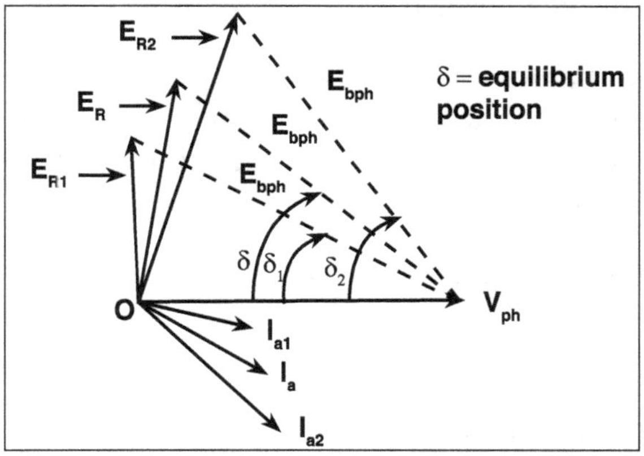

Current variations during hunting.

If such oscillations continue for longer period, there are large fluctuations in the current. So motor gets subjected to large mechanical and electrical stresses. Hence hunting is not a desirable phenomenon from motor point of view and must be prevented.

Two techniques should be used to reduce hunting are:

Use of Damper Winding: It consists of low electrical resistance copper/aluminium brush embedded in slots of pole faces in salient pole machine. Damper winding damps out hunting by producing torque opposite to slip of rotor. The magnitude of damping torque is proportional to the slip speed.

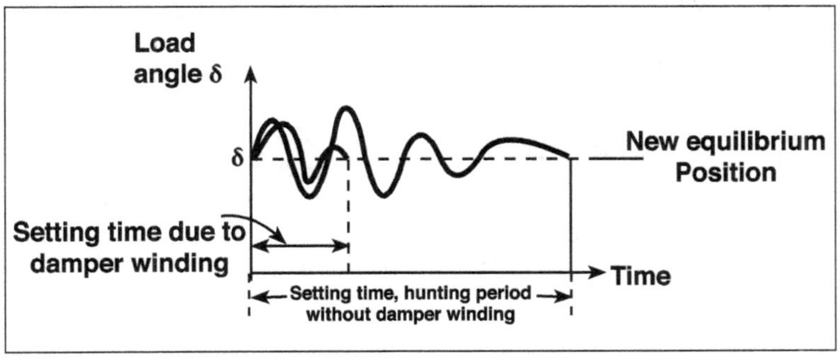

Characteristic curve.

Use of Flywheels

The prime mover is provided with a large and heavy flywheel. This increases the inertia of prime mover and helps in maintaining the rotor speed constant and designing the synchronous machine with suitable synchronizing power coefficients.

Three Phase Synchronous Machines

Constructional Details

Machines which are generating A.C e.m.f are called alternators or synchronous generators whereas, the machines which are accepting the input from the A.C supply to produce mechanical output are called synchronous motors. Both these machines work at a specific constant speed called synchronous speed and hence in general called as synchronous machines.

The working principle of alternator is really very simple. It is similar to the basic principle of DC generator. It also depends upon Faraday's law of electromagnetic induction which says the current is induced in the conductor inside a magnetic field when there is a relative motion between that conductor and the magnetic field.

An alternator mainly consists of two parts such as:

Stator

The stator is a stationary part. It consists of a core and the slots to hold the armature winding similar to the armature of a D.C generator. The stator core uses a laminated construction which is built up of special steel stampings insulated from each other with varnish or paper. The laminated construction is basically to lower the eddy current losses.

Generally, choice of material is steel to keep down hysteresis losses. The entire core is fabricated in a frame made of steel plates. The core contains slots on its periphery for housing the armature conductors. Frame does not carry any flux and serves as the support to the core. Ventilation is maintained with the help of holes cast in the frame.

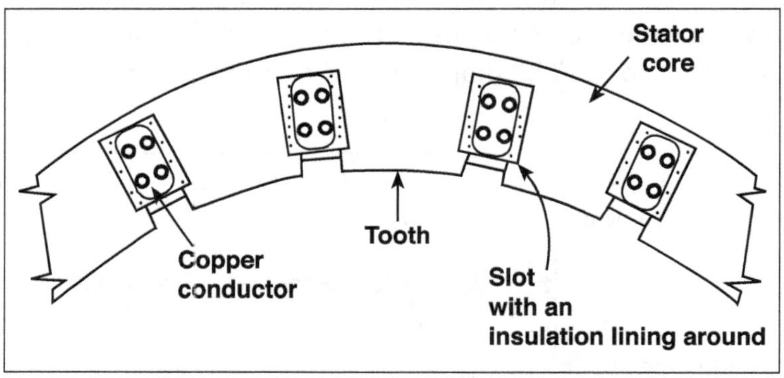

Construction of Stator.

Rotor

There are two types of rotors used in alternators. They are as follows:

Salient Pole Type: This is also known as projected pole type as all the poles are projected out from the surface of the rotor.

The poles are built up of thick steel laminations. They are bolted to the rotor as shown in the above figure. The pole face has been given a specific shape. The field winding is provided on the pole shoe. These rotors have large diameters and small axial lengths. The limiting factor for the size of the rotor is the centrifugal force acting on the rotating member of the machine.

As the mechanical strength of the salient pole type is less, this is preferred for low speed alternators ranging from 125 to 500 r.p.m. The prime movers used to drive such rotor are generally water turbines and I.C. engines.

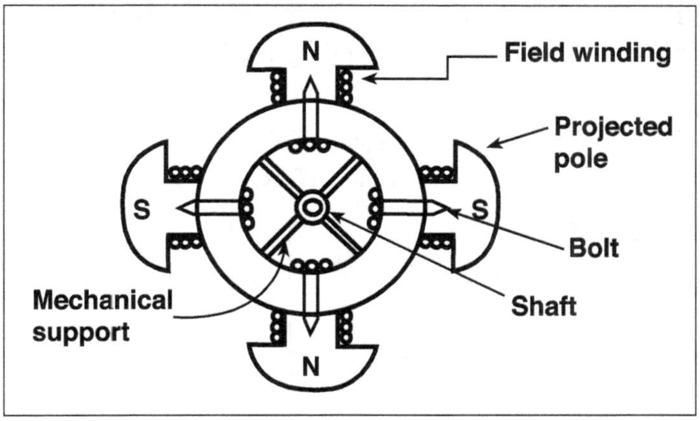

Salient pole type rotor.

Smooth Cylindrical Type: This is also termed as non-salient type or non-projected pole type of rotor. The rotor consists of smooth solid steel cylinder which is having number of slots to accommodate the field coil. The slots are covered at the top with the help of manganese or steel wedges. The unslotted portions of the cylinder itself act as the poles.

The poles are not projecting out and the surface of the rotor is smooth which maintains uniform air gap between stator and the rotor. This rotor has small diameters and large axial lengths. This is to keep peripheral speed within limits. The main advantage of this type is that these are mechanically very strong and thus preferred for high speed alternators ranging between 1500 to 3000 r.p.m. Such high speed alternators are called 'turbo alternators'. The prime movers used to drive such type of rotors are commonly steam turbines, electric motors.

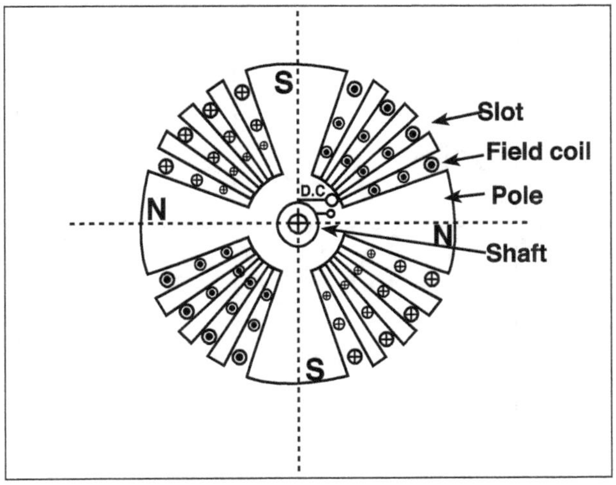

Smooth Cylindrical rotor.

Working Principle

The alternator work on the principle of electromagnetic induction. When there is a relative motion between the conductors and the flux, emf gets induced in the conductors. DC generator also works on the same principle. The only difference in practical alternator and a DC generator is that in an alternator the conductors are stationary and field is rotating. But let us consider the relative motion of the conductors with respect to the flux produced by the field winding.

Consider a relative motion of a single conductor under the magnetic field produced by two stationary poles. The magnetic axis of two poles produced by field is vertical which is shown as dotted in the below figure:

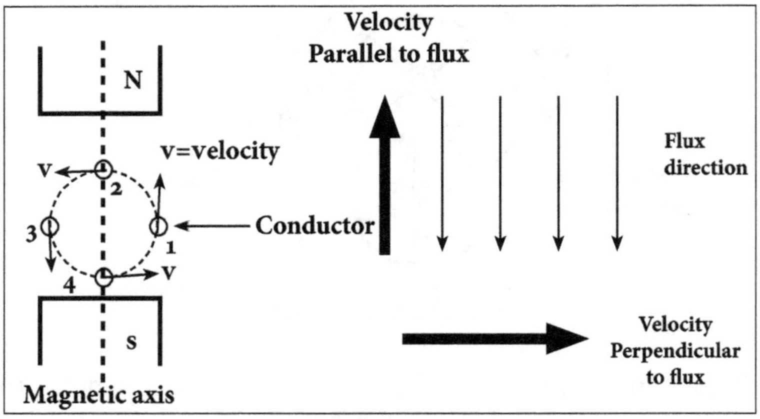

Two pole alternator.

Let the conductor starts rotating from position 1. At this instant, the entire velocity component is parallel to the flux lines. Hence, there is no cutting of flux lines by the conductor. So $d\phi/dt$ at this instant is zero and hence induced emf in the conductor is also zero.

As the conductor moves from position 1 to position 2, the part of the velocity component becomes perpendicular to the flux lines and emf gets induced in the conductor. The magnitude of such an induced emf increases as conductor moves from position 1 to 2.

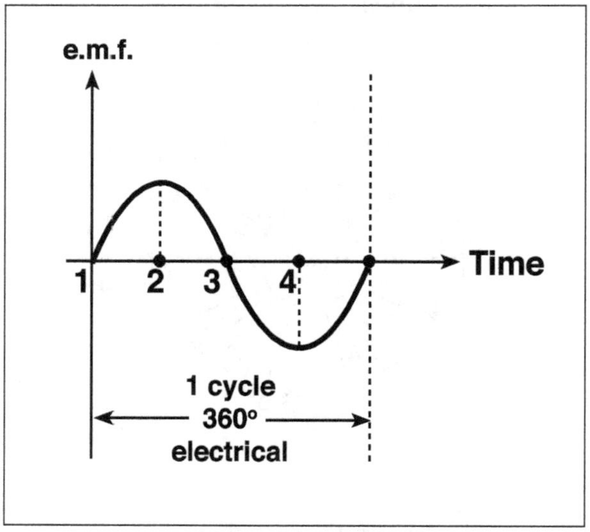

Single cycle of rotation.

At position 2, the entire velocity component is perpendicular to the flux lines. Hence, there exists cutting of the flux lines. At this instant, induced emf in the conductor is maximum. As the position changes from 2 to 3, the velocity component which is perpendicular to the flux starts decreasing and hence induced emf magnitude also starts decreasing.

At position 3, again the entire velocity component is parallel to the flux lines and thus at this instant, induced emf is zero.

As the conductor moves from 3 to 4, the velocity component which is perpendicular to the flux lines again starts increasing. But the direction of velocity component is opposite to the direction of the velocity component existing during the movement of the conductor from position 1 to 2. Thus, the induced emf in the conductor increases but in the opposite direction.

At position 4, it achieves maxima in the opposite direction as the entire velocity component becomes perpendicular to the flux lines. Again from position 4 to 1, induced emf decreases and finally at position 1 it becomes zero. This cycle continues as conductor rotates at a certain speed. Thus, if we plot the magnitude of the induced emf against the time, the alternating nature of the induced emf is shown in the figure above. Now consider 4 pole AC machine and its equivalent circuit as shown in the figure.

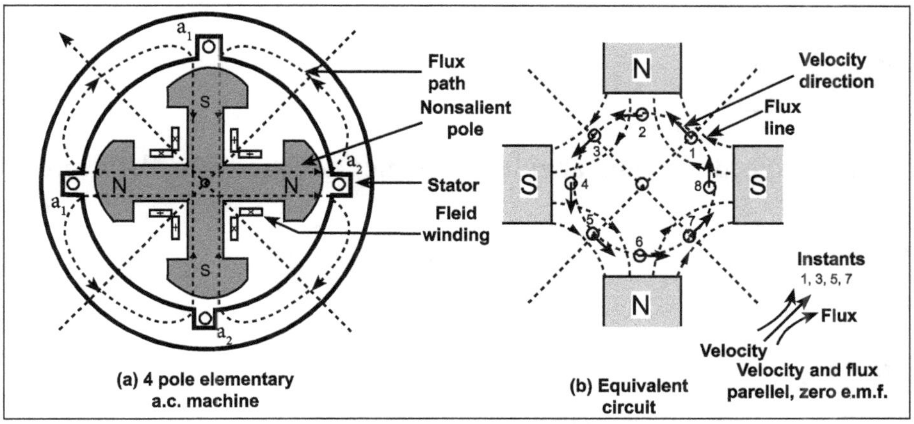

4 pole AC machine and equivalent circuit.

As shown in the above figure (b) at position 1, velocity component of conductor is parallel to the flux lines hence induced emf is zero. In position 2, velocity component cuts maximum flux lines and induced e.m.f is maximum. In position 3, it is again zero as velocity component is parallel to the flux.

In position 4, it is maximum but reversed direction as under opposite pole. This continues as conductor completes one rotation. But now during one mechanical rotation of conductor, induced emf achieves two positive maxima and two negative maxima i.e., two electrical cycles. This is shown in the below figure:

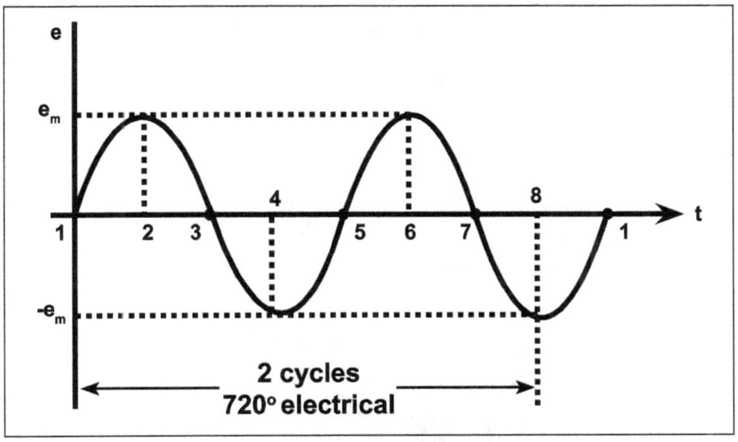

Generating voltage.

Thus for 360° mechanical, 720° electrical are obtained for 4 pole AC machine. Thus it can be written that:

$$\theta_e = \frac{P}{2}\theta_m$$

So in a P pole machine, if speed of rotor is N r.p.m. then it takes one second to complete N/60 revolution. If electrically the frequency is f Hz i.e. emf completes f cycles in one second electrically. And such two cycles in one second are related through P/2.

$$f(\text{electrical}) = \left(\frac{P}{2}\right)\frac{N}{60}(\text{mechanical})$$

$$N = \frac{120f}{P} \text{ or } f = \frac{NP}{120}$$

In angular mechanical speed cum and angular frequency, ω in rad/sec, the expression can be expressed as:

$$\omega = \left(\frac{P}{2}\right)\omega_m$$

It is known that:

$$\omega = 2\pi f \text{ rad/sec} \text{ ...Electrical relation}$$

The speed at which AC machine generates a voltage at the rated frequency which is 50 Hz is called the synchronous speed denoted as N_S.

$$N_s = \frac{120f}{P} \text{ rpm.}$$

Synchronous Motor

Construction

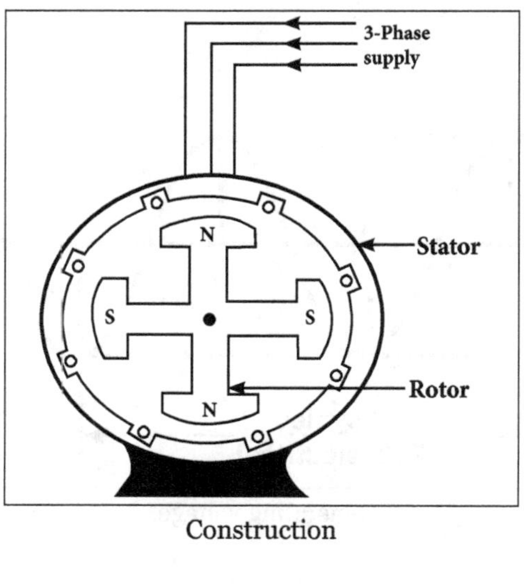

Construction

A synchronous motor is a machine which operates at synchronous speed and converts electrical energy into mechanical energy. It is basically an alternator operated as a motor. Like an alternator, a synchronous motor has the following two parts:

- Stator which houses 3-phase armature winding in the slots of stator core and receives power from a 3-phase supply.
- Rotor that has a set of salient poles excited by direct current to form alternate N and S poles. The exciting coils are connected in series to two slip rings and direct current is fed into the winding from an external exciter mounted on rotor shaft. The stator is wound for the same number of poles as the rotor poles.

As in the case of an induction motor, the number of poles determines the synchronous speed of the motor:

$$N_s = 120f / P$$

Where,

P = Number of poles

f = Frequency of supply in Hz

An important disadvantage of a synchronous motor is that it is not self-starting and auxiliary means have to be used for starting it.

Principle of Operation

(a) Consider a 3-phase synchronous motor having two rotor poles N_R and S_R. Then the stator will be wound for two poles N_S and S_S. The motor has direct voltage applied to the rotor winding and a 3-phase voltage applied to the stator winding. The stator winding produce a rotating field which revolves round the stator at synchronous speed N_s.

$$N_s = 120f / P$$

The direct current sets up a two-pole field which is stationary as lng as the rotor is not turning. Therefore, we have a situation in which there exists a pair of revolving armature poles (i.e., N_S - S_S) and a pair of stationary rotor poles (i.e., N_R - S_R).

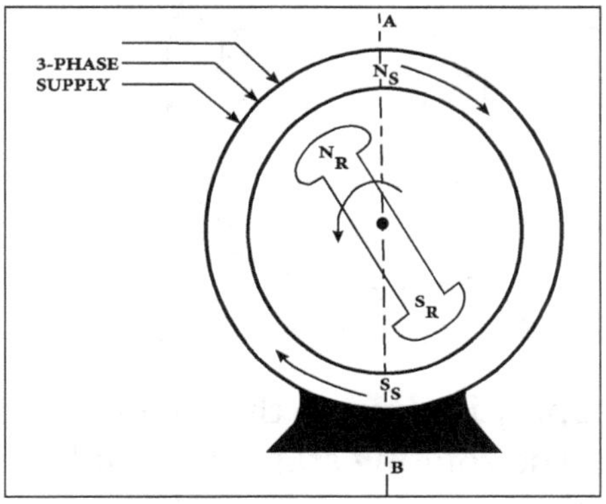

(a) Synchronous motor.

(b) Suppose at any instant, the stator poles are at positions A and B as shown in Figure (a). It is clear that poles N_S and N_R repel each other and so do the poles S_S and S_R. Therefore, the rotor tends to move in the anticlockwise direction.

After a period of half-cycle (½ f = 1/100 second), the polarities of the stator poles are reversed but the polarities of the rotor poles remain the same as shown in Figure (b). Now S_S and N_R attract each other and so do N_S and S_R. Thus, the rotor tends to move in the clockwise direction. Because the stator poles change their polarities rapidly, they tend to pull the rotor first in one direction and then after a period of half-cycle in the other. Due to high inertia of the rotor, the motor fails to start.

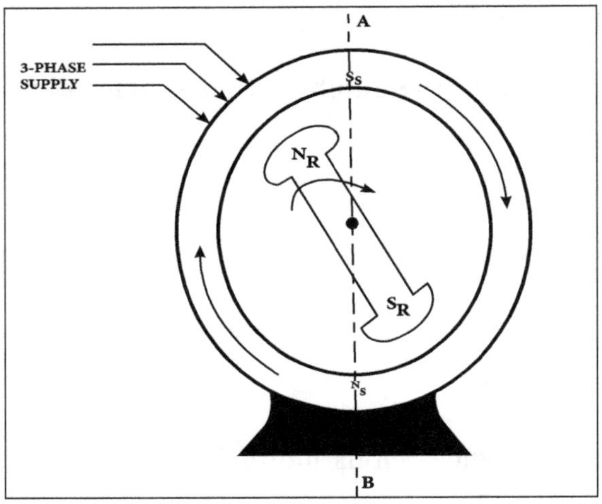

(b) Synchronous motor.

Hence, a synchronous motor has no self-starting torque i.e., a synchronous motor cannot start by itself.

Making Synchronous Motor Self-Starting

The synchronous motor cannot start by itself. In order to make the motor self-starting, a squirrel cage winding is provided on the rotor. The damper winding consists of copper bars embedded in the pole faces of the salient poles of the rotor as shown in the below figure. The bars are short-circuited at the ends to form a partial squirrel cage winding. The damper winding serves to start the motor.

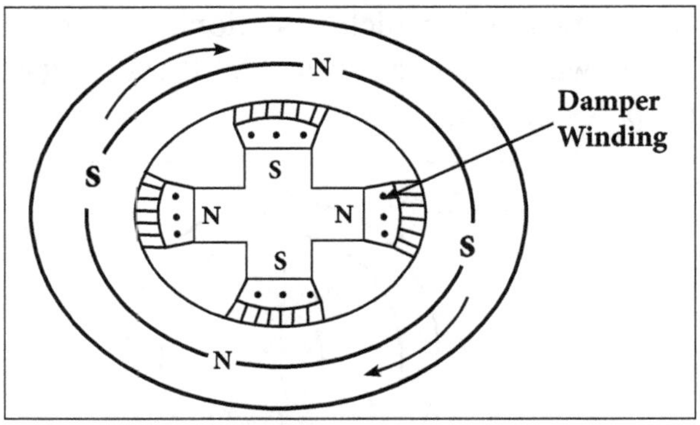

Salient pole synchronous motor.

- To start with, 3-phase supply is given to the stator winding while the rotor field winding is left un-energized. The rotating stator field induces currents in the squirrel cage or damper winding and the motor starts as an induction motor.

- As the motor approaches the synchronous speed, the rotor is excited with direct current. Now the resulting poles on the rotor face poles of opposite polarity on the stator and a strong magnetic attraction is set up between them. The rotor poles lock in with the poles of rotating flux. Subsequently, the rotor revolves at the same speed as the stator field i.e., at synchronous speed.
- Because the bars of squirrel cage portion of the rotor now rotate at the same speed as the rotating stator field, these bars do not cut any flux and therefore, have no induced currents in them. Therefore squirrel cage portion of the rotor is, in effect, removed from the operation of the motor.

It can be emphasized here that due to magnetic interlocking between the stator and rotor poles, a synchronous motor only run at synchronous speed. At any other speed, this magnetic interlocking (i.e., rotor poles facing opposite polarity stator poles) ceases and the average torque becomes zero. Therefore, the motor comes to a halt with a severe disturbance on the line.

It is important to excite the rotor with direct current at the right moment. For example, if the DC excitation is applied when N-pole of the stator faces N pole of the rotor, the resulting magnetic repulsion will produce a violent mechanical shock. The motor can immediately slow down and the circuit breakers will trip. In practice, starters for synchronous motors are designed to detect the precise moment when excitation should be applied.

1.2.1 3-phase Induction Machine

Principle of Operation

An electrical motor is an electromechanical device which converts electrical energy into a mechanical energy. In case of three phase AC operation, most widely used motor is three phase induction motor as this type of motor does not require any starting device or we can say, they are self-starting induction motor. The stator of the motor consists of overlapping winding offset by an electrical angle of 120°. When the primary winding of the stator is connected to the 3 phase AC source, it establishes a rotating magnetic field which rotates at the synchronous speed.

According to Faraday's law, an emf induced in any circuit is due to the rate of change of magnetic flux linkage through the circuit. As the rotor winding in an induction motor are either closed through an external resistance or directly shorted by the end ring and cut the stator rotating magnetic field, an emf is induced in the rotor copper bar and due to this emf, current flows through the rotor conductor. Here, the relative speed between the rotating flux and static rotor conductor is the cause of current generation.

According to Lenz's law, the rotor will rotate in the same direction to reduce the cause. i.e., the relative velocity. Thus from the working principle of three phase induction motor, it is observed that the rotor speed should not reach the synchronous speed produced by the stator.

If the speeds equals, there could be no such relative speed, so no emf induced in the rotor and no current could be flowing and therefore, no torque could be generated. Consequently, the rotor cannot reach the synchronous speed. The difference between the stator and rotor speeds is called the slip. The rotation of the magnetic field in an induction motor has the advantage that no electrical connections need to be made to the rotor.

The rotating magnetic field cuts the rotor windings and produces an induced voltage in the rotor windings. Because of the fact that the rotor windings are short circuited for squirrel cage and wound-rotor and induced current flows in the rotor windings, the rotor current produces another magnetic field.

A torque is produced as a result of the interaction of those two magnetic fields:

$$\tau_{ind} = kB_R \text{ X } B_s$$

Where,

τ_{ind} - The induced torque

B_R and B_s are the magnetic flux densities of the rotor and the stator respectively.

IM can run at the synchronous speed because, if rotor runs at the synchronous speed, which is the same speed of the rotating magnetic field, then the rotor will appear stationary to the rotating magnetic field and the rotating magnetic field will not cut the rotor.

No induced current will flow in the rotor and no rotor magnetic flux will be produced. No torque is generated and the rotor speed will fall below the synchronous speed. Hence, the IM will always run at a speed lower than the synchronous speed.

Construction

The three phase induction motor is the most commonly used electrical motor. Almost 80% of the mechanical power used by industries is provided by three phase induction motors due to its simple and rugged construction, absence of commutator and good speed regulation, low cost and good operating characteristics.

In three phase induction motor, the power is transferred from stator to the rotor winding through induction. The induction motor is also known as a synchronous motor as it runs at a speed other than the synchronous speed. Three phase induction motor has two main parts. They are,

Stator

Stator consists of a steel frame that supports a hollow and cylindrical core of stacked laminations. Slots on the internal circumference of the stator poles carry the stator winding. It is a stationary part of the induction motor.

A stator winding is placed in the stator of induction motor and the three phase supply is given to it. The stator of the three phase induction motor consists of three main parts such as:

- Stator frame.
- Stator core.
- Stator winding or field winding.

Depending on the type of rotor construction used, the three phase inductor motor are classified as,

- Squirrel cage induction motor.
- Slip ring induction motor or wound induction motor or phase wound induction motor.

The construction of stator for both the types of three phase induction motor remains the same.

- There is a small distance between rotor and the stator which normally varies from 0.4 mm to 4 mm. Such a distance is termed as air gap.
- Shaft for transmitting the torque to the load. This shaft is made up of steel.

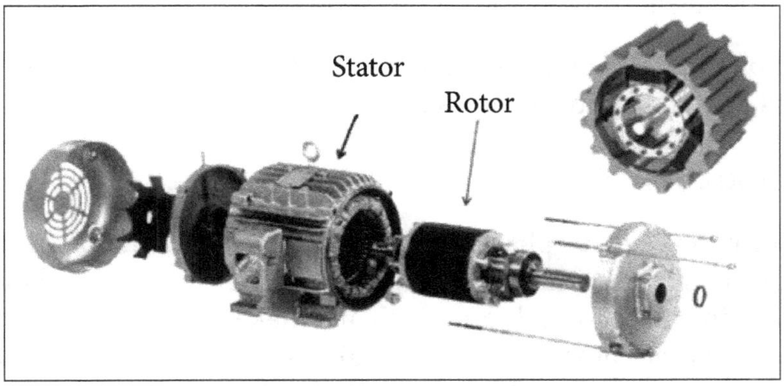

Cross sectional view of 3-phase induction motors.

- One of the problems with electrical motor is the production of heat during its rotation. In order to overcome this problem we need fan for cooling.
- Bearings for supporting the rotating shaft.
- For receiving external electrical connection terminal box is needed.

Stator Frame

It is the outer most part of the three phase induction motor. Its main function is to support the stator core and the field winding. It acts as a covering and it provide protection and mechanical strength to all the inner parts of the induction motor. The frame is either made up of die cast or fabricated steel.

The frame of three phase induction motor could be very strong and rigid as the air gap length of three phase induction motor is quite small. Otherwise, rotor cannot remain concentric with stator, which will give rise to unbalanced magnetic pull.

Stator Core

The main function of the stator core is to carry the alternating flux. In order to reduce the eddy current loss, the stator core is laminated. These laminated types of structure are made up of stamping which is about 0.4 to 0.5 mm thick.

All the stampings are stamped together to form a stator core, which is then housed in the stator frame. Stamping is commonly made up of silicon steel, which helps to reduce the hysteresis loss occurring in motor.

Stator Winding or Field Winding

The slots on the periphery of stator core of the three phase induction motor carries three phase windings. This three phase winding is supplied by the three phase AC supply. The three phases of the winding are connected either in star or delta depending on which type of starting method is used.

The squirrel cage motor is started by star-delta starter and hence the stator of squirrel cage motor is delta connected. The slip ring three phase induction motor are started by inserting resistances because, the stator winding of slip ring induction motor will be connected either in star or delta.

The winding wound on the stator of three phase induction motor is also called as field winding and when this winding is excited by three phase AC supply, it produces a rotating magnetic field.

Squirrel Cage Induction Motor

The rotor of the squirrel cage three phase inductions motor is cylindrical in shape and has slots on its periphery. The slots are not made parallel to each other but are bit skewed as the skewing prevents magnetic locking of stator and rotor teeth and makes the working of motor more smooth and quieter. The squirrel cage rotor consists of aluminum, brass. This aluminum, brass or copper bars are known as rotor conductors and are placed in the slots on the periphery of the rotor.

The rotor conductors are permanently shorted by the copper or aluminum rings known as the end rings. In order to provide mechanical strength these rotor conductor are braced to the end ring and therefore form a complete closed circuit resembling like a cage and so got its name as "squirrel cage induction motor".

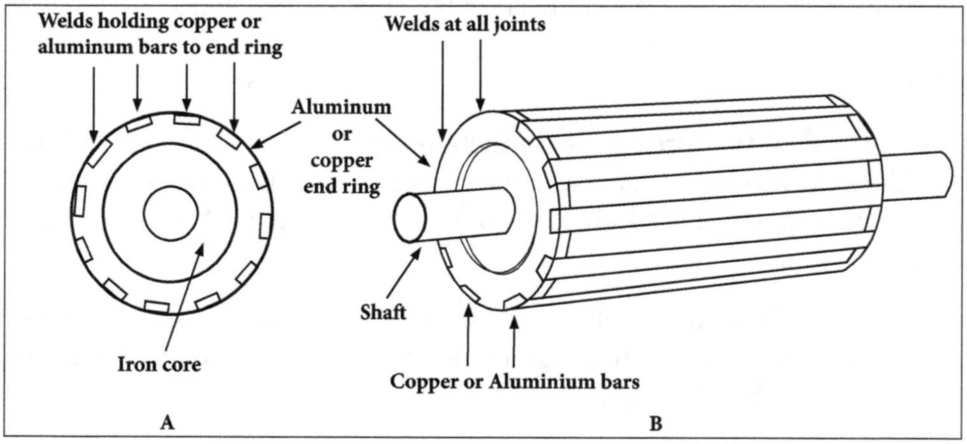

Squirrel cage rotor for induction motor.

The squirrel cage rotor winding is made symmetrical. As the bars are permanently shorted by end rings, the rotor resistance is quite small and it is not possible to add external resistance as the bars are permanently shorted. The absence of slip ring and brushes make the construction of the Squirrel cage three phase induction motor quite simple and robust and hence, it is the widely used three phase induction motor. These motors have the advantage of adapting any number of pole pairs. The above diagram shows squirrel cage induction rotor having aluminum bars short circuit by aluminum end rings.

Advantages

- As there are no brushes and slip ring, these motors requires less maintenance.
- Its construction is very simple and rugged.

Applications

Squirrel cage induction motor is used in lathes, fan, drilling machine, blower printing machines etc.

Wound Rotor or Slip Ring Induction Motor

In this type of motor, the rotor is wound for the same number of poles as that of stator but it has less number of slots and has less turns per phase of a heavier conductor. The rotor also carries the star or delta winding similar to that of the stator winding. The rotor consists of numbers of slots and rotor winding are placed inside these slots.

The three end terminals are connected together to form star connection. As its name indicates, the three phase slip ring induction motor consists of slip rings connected on same shaft as that of rotor. The three ends of the three phase windings are permanently connected to these slip rings.

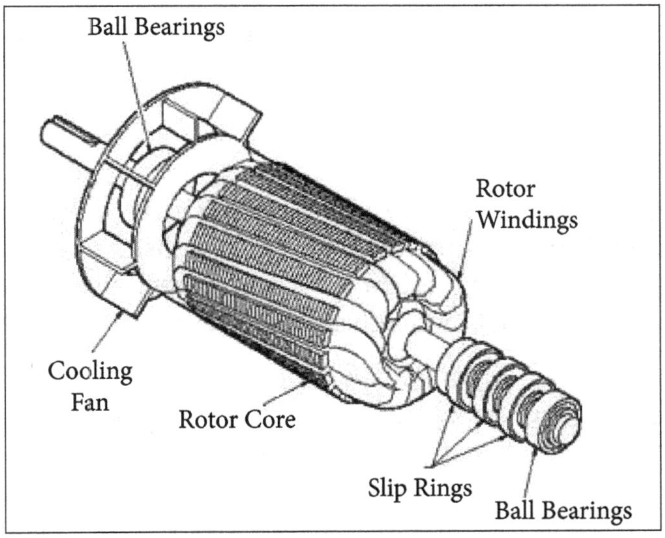

Cross-section view of ring rotor.

The external resistance can be easily connected through the brushes and slip rings. It is used for speed control and improving the starting torque of three phase induction motor. The brushes are used to carry current to and from the rotor winding. These brushes are further connected to the three phase star connected resistances.

At starting, the resistances are connected in rotor circuit and are gradually cutout as the rotor pick up its speed. When the motor is running the slip ring is shorted by connecting a metal collar, that connect all slip ring together and the brushes are also removed. This reduces wear and tear of the brushes. Because of the presence of slip rings and brushes, the rotor construction gets somewhat complicated hence it is less used as compared to squirrel cage induction motor.

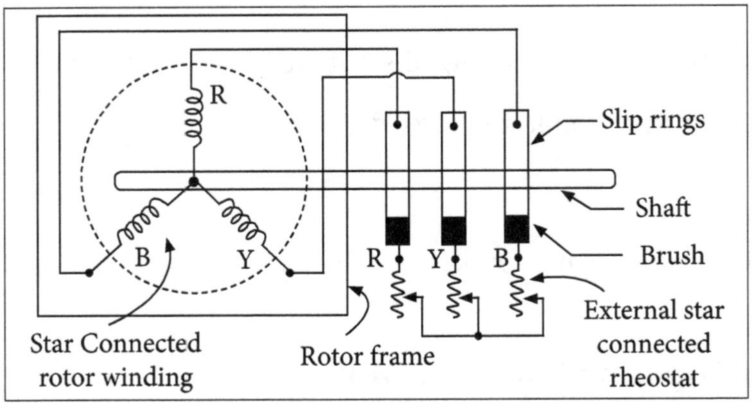

Wound rotor induction motor.

Advantages

- It has high starting torque and low starting current.
- Possibility of adding additional resistance to control speed.

Application

Slip ring induction motor are used where high starting torque is required i.e., in hoists, cranes, elevator etc.

Equivalent Circuit

The equivalent circuit of any machine shows different parameters of the machine such as its ohmic losses and also other losses. The losses are modeled just by resistor and inductor. The copper losses occur in the windings, so the winding resistance is considered. Also, the winding has inductance for which there is a voltage drop due to inductive reactance.

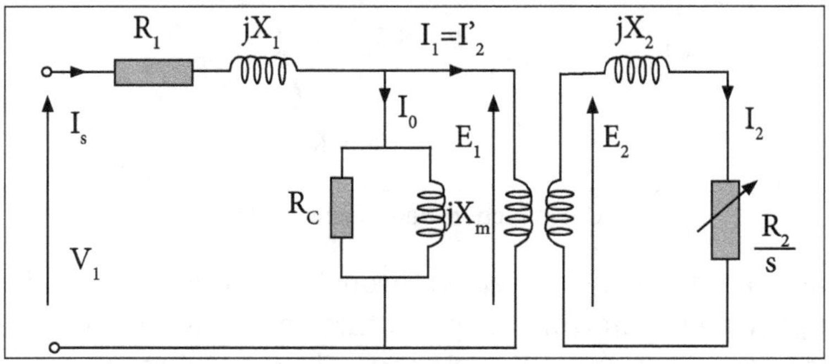

Equivalent circuit.

Here,

R_1 is the winding resistance of the stator.

X_1 is the inductance of the stator winding.

R_c is the core loss component.

X_M is the magnetizing reactance of the winding.

R_2/s is the power of the rotor, which includes output mechanical power and copper loss of rotor. Below figure shows the circuit with reference to the stator.

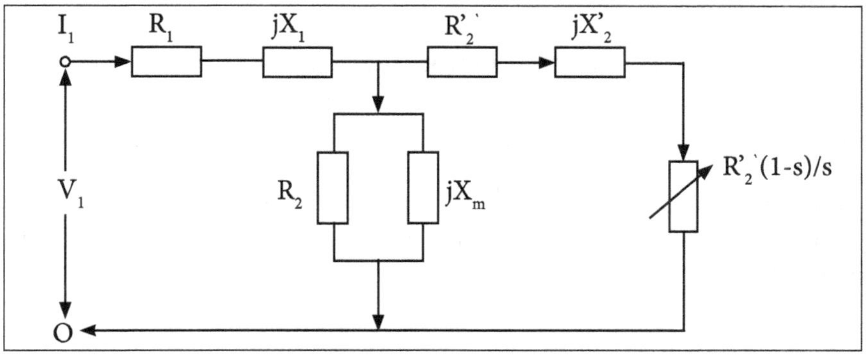

Equivalent circuit.

Here all the other parameters are same except R_2', is the rotor winding resistance with reference to stator winding. X_2' is the rotor winding inductance with reference to stator winding. $R_2(1 - s)/s$ is the resistance which shows the power which is converted into mechanical power output or useful power. The power dissipated in that resistor is the useful power output or shaft power.

Approximate Equivalent Circuit

The approximate equivalent circuit is drawn just to simplify our calculation by deleting one node. The shunt branch is shifted towards the primary side. This has been done as the voltage drop between the stator resistance and inductance is less and the difference between the supply voltage and the induced voltage is not much. Although, this is not appropriate due to following reasons:

- The magnetic circuit of induction motor has an air gap, so exciting current is large when compared to transformer. Therefore, exact equivalent circuit should be used.
- The rotor and stator inductance is large in induction motor.
- In induction motor, we use distributed windings. This model will be used if approximate analysis has to be done for large motors. This model will not be used for the analysis of small motors.

Power Relation of Equivalent Circuit

Input power to stator = $3\ V_1 I_1 \cos\theta$

Where,

- V_1 is the stator voltage applied.
- I_1 is the current drawn by the stator winding.

- Cos(θ) is the stator power factor.
  - Rotor input = Power input - Stator copper and iron losses.
  - Rotor copper loss = Slip × Power input to the rotor.
  - Developed power = (1 - s) × Rotor input power.

Phasor Diagram

The phasor diagram of a loaded induction motor is similar to the loaded transformer. The only difference is the secondary of induction motor is rotating and short circuited whereas, transformer secondary is stationary and connected to the load.

The load on the induction motor is mechanical whereas, the load on the transformer is electrical. Still by finding electrical equivalent of mechanical load on the motor, the phasor diagram of induction motor can be developed.

Let,

φ = Magnetic flux links with both primary and secondary.

There is self-induced emf E_1 in the stator while a mutually induced emf E_2 in the rotor.

Let,

R_1 = Stator resistance per phase

X_1 = Stator reactance per phase

The stator voltage per phase V_1 has to counter balance the self-induced emf E_1 and has to supply voltage drops I_1R_1 and I_1X_1. So on stator side, we can write:

$$\overline{V_1} = -\overline{E_1} + \overline{I_1R_1} + j\overline{I_1X_1} = \overline{E_1} + \overline{I_1}(R_1 + jX_1) = -\bar{E}_1 + \bar{I}_1 + \bar{Z}_1$$

The rotor induced emf in the running condition has to supply the drop across impedances as rotor is short circuited. The value of E_{2r} depends on the ratio of rotor turns to stator turns. The rotor current in the running condition is I_{2r} which lags E_{2r} by rotor pf angle φ_{2r}.

The reflected rotor current I_{2r}' on stator side is the effect of load and is given by:

$$I_{2r}' = KI_{2_r}$$

The induction motor draws no load current I_0 which is phasor sum of I_c and I_m. The total stator current drawn from supply is given by:

$$\hat{I}_1 = \hat{I}_0 + \hat{I}_{2r}'$$

Where φ_1 is the angle between V_1 and I_1 and cos φ_1 gives the power factor of the induction motor. Thus, using all the above relations, the phasor diagram of induction motor on load can be obtained.

Steps:

The steps to draw phasor diagram are as follows:

- Takes φ as reference phasor.
- The induced voltage E_1 lags φ by 90°.
- Show $-E_1$ by reversing voltage phasor.
- The phasor E_{2r} is in phase with E_1. So I_{2r} lags E_{2r} i.e. E_1 direction by φ_{2r}.
- Show I_{2r} R_2 in phase with I_{2r} and I_{2r} X_{2r} leading the resistive drop by 90°, to get the exact location.
- Reverse I_{2r} to get I_{2r}'.
- I_m is in phase with φ while I_c is at leading. Add I_m and I_c to get I_o.
- Add I_o and I_{2r}' to get I_1.
- From tip of - E_1 phasor, add I_1 R_1 in phase with I_1 and I_1X_1 at 90° leading to I_1 to V_1 get phasor.
- Angle between V_1 and I_1 is φ_1.The phasor diagram is shown in the figure:

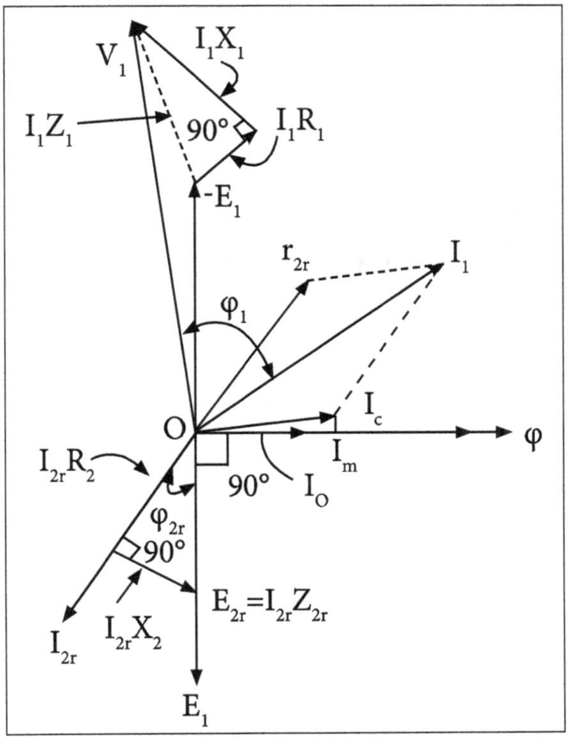

Phasor diagram of induction motor.

Torque Equation of an Induction Motor

We know that total torque developed is:

$$T_m = \left(\frac{3}{\omega_s}\right) \cdot \frac{sE_2^2}{\left[R_2^2 + (sX_2)^2\right]} . R_2$$

Let us examine the torque verses speed characteristics for different operating conditions.

Case 1: Motor running near synchronous speed (s very small).

Case 2: At Starting, (s=1).

Case 1: Motor running near synchronous speed:

General torque expression is given by,

$$T_m = \left(\frac{3}{\omega_s}\right) \cdot \frac{sE_2^2}{\left[R_2^2 + (sX_2)^2\right]} . R_2$$

Slip s is very small and hence R22 >> (sx2)2. So torque expression becomes:

$$T_m = \left(\frac{3}{\omega_s}\right) \cdot \frac{sE_2^2}{R_2}$$

Where α denotes proportionality:

$$T_m \, \alpha \, \frac{sE_2^2}{R_2}$$

Near Synchronous Speed

Torque increases linearly with slip. If rotor resistance is high, rated torque is reduced. Torque is proportional to the square of applied voltage.

Case 2: At starting:

General torque expression is:

$$T_m = \left(\frac{3}{\omega_s}\right) \cdot \frac{E_2^2}{X_2^2} . R_2$$

$$T_{st} \, \alpha \, \frac{E_2^2 . R_2}{X_2^2}$$

Where α denotes proportionality.

During Starting

Starting Torque increases linearly with rotor resistance. If leakage reactance is high, starting torque is reduced. Torque is proportional to the square of applied voltage.

1.3 Kron's Primitive Machine

The Kron's primitive machine with stationary axes consists of a stationary field winding DS in the direct axis, an independent field winding QS at right angles to d-axis, i.e., in the quadrature axis and a rotating armature winding brought out to a commutator. Two sets of brushes are provided which are magnetically perpendicular to each other, with one brush set in the d-axis and the other in the q-axis. The Kron's primitive machine with stationary axes is as shown in figure I(a).

The idealized or basic two-pole machine diagram for this machine is shown in figure I(b). This equivalent electrical network is termed as "generalized machine", "Kron's primitive machine", "generalized model or two-axis model" of rotating electrical machines. It can be noted that the equivalent armature coils DR and QR produce stationary fluxes along the brush axes, only if the brushes are stationary. If the brushes are rotating, the corresponding armature coil fluxes will also be rotating.

Single armature winding is made to act as two electrically and magnetically independent armature windings in quadrature, due to the presence of two brush sets. This generalized primitive machine can be shown to be equivalent to any of the rotating electrical machines with an appropriate number of coils on each fixed axis. Some machines may require fewer than four coils to represent them, while others may require more.

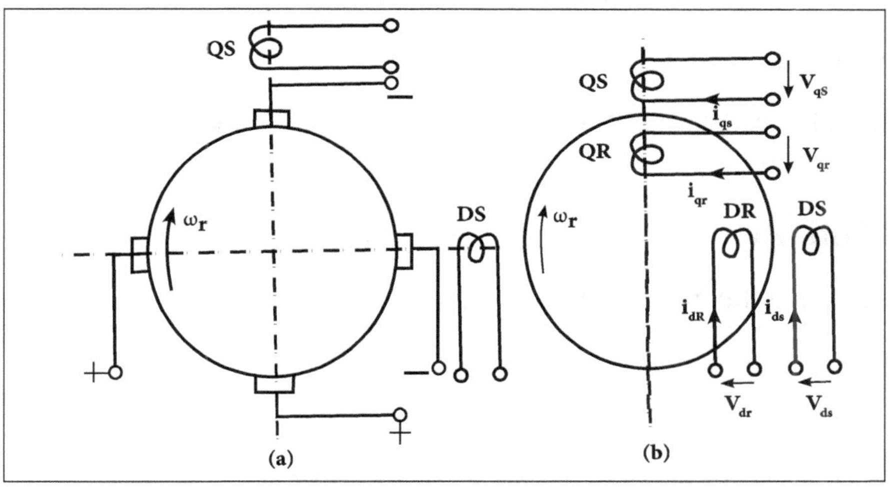

I. (a) Kron's primitive machine and (b) Its basic two-pole representation.

For the ordinary d.c. machine, the generalized machine diagram requires only two coils, DS for the field winding and QR for the armature. Thus, the coils DR and QS will have to be omitted from figure (b) to obtain a generalized machine diagram for this machine.

Some of the basic two-pole machines are similar to the generalized or primitive machine, but if they are not, it is necessary to make a conversion. The process of conversion from the actual coils of the machine itself to the equivalent d and q-axes of the generalized machine or vice-versa is known as a transformation.

For instance, in the case of a three-phase winding on the stator, a transformation must be made to the equivalent coils DS and QS, in such a manner that the magnitude and direction of m.m.f. set up by the three-phase currents i_a, i_b, i_c and the two coil currents i_{ds}, i_{qs} is identical.

The rotor of this machine should not know whether the rotating m.m.f. is produced by three phase currents i_a, i_b, i_c or by two coil currents i_{ds}, i_{qs}. Thus, the generalized machine diagram for a 3-phase induction machine is as shown in the below figure II (a). Similarly, the primitive machine diagram for a synchronous machine (without damper bars) is as shown in figure II (b).

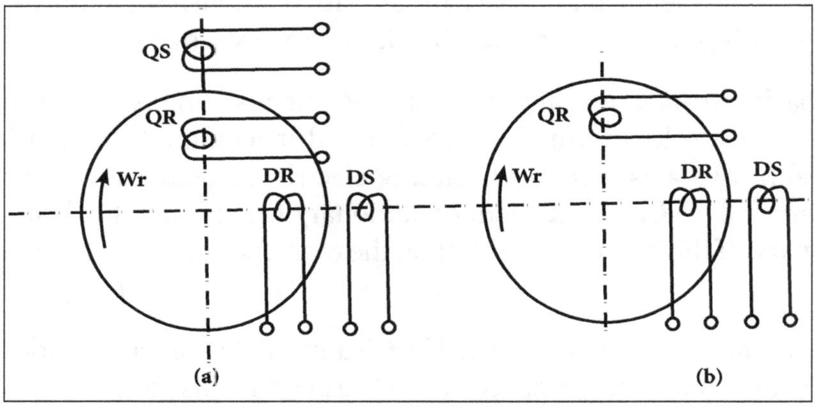

II. Primitive (or generalized) machine diagrams (a) Induction machine (b) Synchronous machine (without dampers).

1.3.1 Voltage, Current and Torque Equations

Voltage Equations

At this stage, the voltage equations for the primitive machine of figure I(b), can be written as follows:

Stator Field Coil DS

Its resistance is r_{ds} and total inductance is given by:

$$L_{ds} = (M_d + l_{ds})$$

Where,

L_{ds} is self-inductance of the coil DS.

M_d is the mutual inductance between the coils DR and DS.

l_{ds} is the leakage inductance of coil DS.

It is mutually coupled with any other d-axis coil, i.e., when the voltage equation for the DS circuit is to be written, the mutual effect of other d-axis coils must be considered. Q-axis coils can have no effect on DS and no rotational voltages appear in it, since the coil DS is on the stationary element. Thus, the applied voltage v_{ds} must account for the components given by:

$$v_{ds} = r_{ds}i_{ds} + L_{ds}pi_{ds} + M_d pi_{dr} \quad ...(1)$$

Stator Field Coil QS

The coil QS is identical with coil DS, so that applied voltage v_{qs} is given by,

$$v_{qs} = r_{qs}i_{qs} + L_{qs}pi_{qs} + M_q pi_{qr} \quad ...(2)$$

Armature Coil DR

The armature coil DR and QR are the pseudo-stationary coils and they have rotational voltages induced in them.

$$v_{dr} = r_{dr}i_{dr} + L_{dr}pi_{dr} + M_d pi_{ds} - e_{dr}$$

Where e_{dr} is the induced rotational voltage.

The negative sign before the induced rotational voltage in the above equation is due to the fact that, if other voltage drops are neglected, the induced voltage is equal and opposite to the applied voltage.

Now using $e_d = +\omega_r \psi_{mq}$ and noting that the rotational emf in coil DR appears due to q-axis flux, we get:

$$e_{dr} = +\omega_r \psi_q$$

In the rotational emf relation e_{dr}, note that the total armature flux linkage ψ_θ is taken into consideration as assumed before.

Substituting the value of e_{dr} in the expression of v_{dr} gives:

$$v_{dr} = r_{dr}i_{dr} + L_{dr}pi_{dr} + M_d pi_{ds} - \omega_r \psi_q$$

Here $\omega_r \psi_q$, the total flux linkages with the armature in q axis, is given by:

$$\psi_q = M_q\left(i_{qs} + i_{qr}\right) + l_{qr}iqr$$

Thus,

$$v_{dr} = r_{dr}i_{dr} + L_{dr}pi_{dr} + M_d pi_{ds} - \omega_r M_q\left(i_{qs} + i_{qr}\right) - \omega_r l_{qr}i_{qr}$$

$$= r_{dr}i_{dr} + L_{dr}pi_{dr} + M_d pi_{ds} - \omega_r L_{qr}i_{qr} - \omega_r M_q i_{qs} \quad ...(3)$$

Armature Coil QR

This armature coil QR is similar to coil DR and therefore adopting the above procedure, the applied voltage v_{qr} must have the components.

$$v_{qr} = r_{qr} i_{qr} + L_{qr} p i_{qr} + M_q p i_{qs} + \omega_r L_{dr} i_{dr} + \omega_r M_d i_{ds} \qquad ...(4)$$

The above four equations from (1) to (4) can be written down rapidly in matrix form as illustrated below:

			ds	qs	dr	qr
v_{ds}		ds	$rd_s + L_{ds}P$		M_{dp}	
v_{qs}	=	qs		$r_{qs} + L_{qs}p$		$M_q p$
v_{dr}		dr	$M_d p$	$-M_q \omega_r$	$r_{dr} + L_{dr}p$	$-\omega_r L_{qs}$
v_{qr}		qr	$\omega_r M_d$	$M_q p$	$\omega_r L_{dr}$	$r_{qr} + L_{qr}p$

Power-Angle Curve

A synchronous machine supplies an electric network with constant voltage under steady state conditions. The terminal voltage in the machine is kept constant by the regulation of field current. The generator speed is constant at the synchronous speed determined by the network frequency and the number of poles in the machine. An increase of input mechanical power increases the torque.

Let us calculate the output power variation with the input power. The external network is represented by a voltage source and an equivalent reactance. A large network's impedance is very small, we assume $X_e = 0$ and $V_{en} = V_{tn}$.

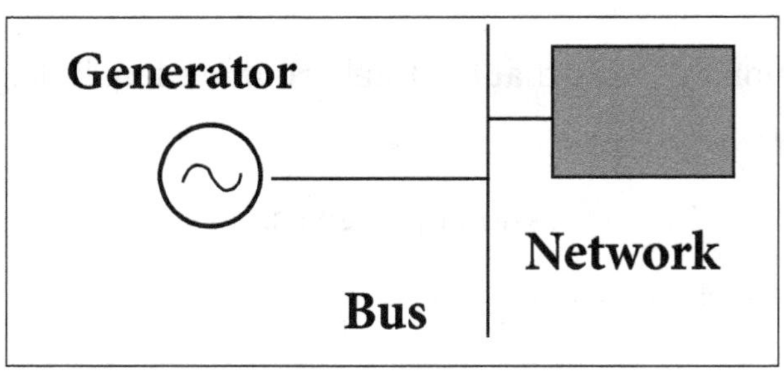

Single line representation.

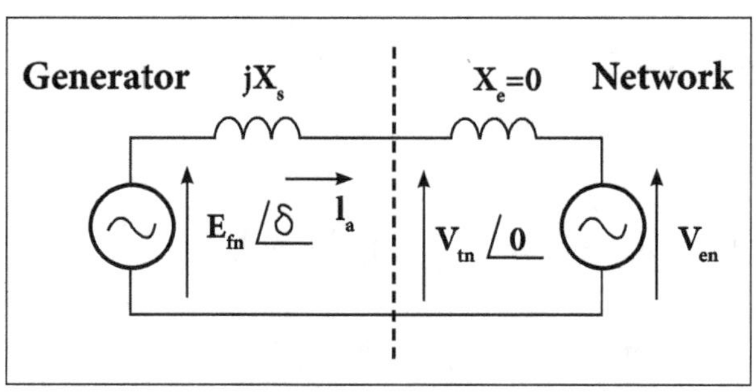

Equivalent circuit.

Using the equivalent circuit, the current is given by:

$$I = \frac{E_{fn}\, e^{i\delta} - V_{tn}}{i X_{syn} + i X_{nt}}$$

The complex power delivered by the generator is given by:

$$S = 3V_{tn}\, \overline{I} = 3\,V_{tn}\left[\frac{E_{fn} e^{-i\delta} - V_{tn}}{-i X_s}\right] \quad X_s = X_{syn} + X_{nt}$$

After simplification, we get:

$$S = 3\frac{E_{fn}\cdot V_{tn}}{X_s}\cdot \sin\delta + j\cdot 3\cdot\left[\frac{E_{fn} V_{tn}}{X_s}\cdot\cos\delta - \frac{V_{tn}^{\ 2}}{X_s}\right]$$

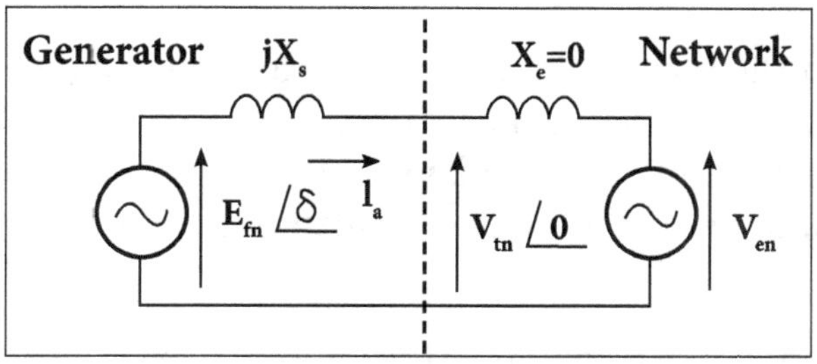

Equivalent circuit.

The real and reactive power are expressed as:

$$P = 3\frac{E_{fn}\cdot V_{tn}}{X_s}\cdot\sin\delta$$

$$Q = j\cdot 3\cdot\left[\frac{E_{fn}\cdot V_{tn}}{X_s}\cdot\cos\delta - \frac{V_{tn}^{\ 2}}{X_s}\right]$$

The real power is maximum if δ = 90°.

The maximum torque is expressed as:

$$T_{max} = P_{max}/\omega$$

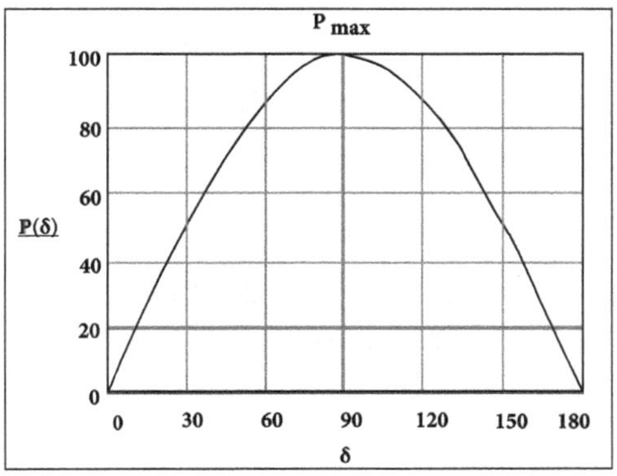

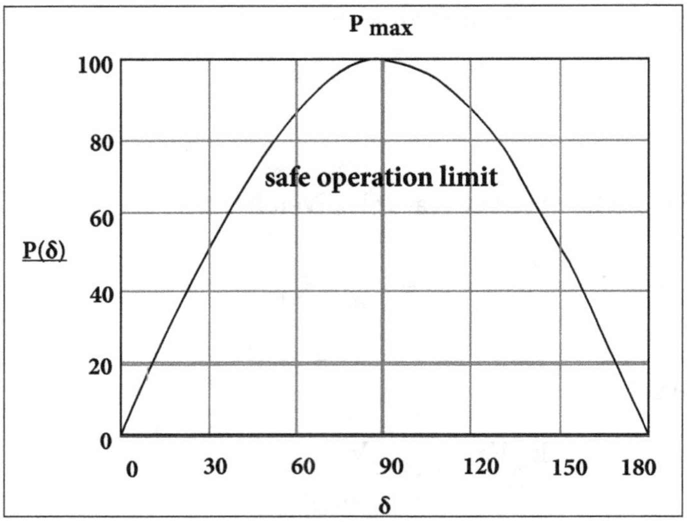

Power angle Characteristics.

The P(δ) curve shows that the increase of power increases the angle between the induced voltage and the terminal voltage. The power is maximum when $\delta = 90°$. Further increase of input power forces the generator out of synchronism. This generates large current and mechanical forces. This angle corresponds to the angle between the field flux and the stator generated rotating flux.

The angle δ, called power angle, corresponds to the angle between the field flux and the stator generated rotating flux. The maximum power is the static stability limit of the system. Safe operation requires a 15-20% power reserve.

Torque Expression

When a synchronous machine is operated as a motor to drive a mechanical load, in steady state, the mechanical torque of the motor must balance the load torque and the mechanical loss torque due to friction and windage:

$$T = T_{load} + T_{loss.}$$

Multiplying the synchronous speed to each sides of the torque equation, we have the power balance equation as:

$$P_{em} = P_{load} + P_{loss}$$

Where $P_{em} = T\omega_{syn}$ is the electromagnetic power of the motor, $P_{load} = Tl_{oad}\ \omega_{syn}$ is the mechanical power delivered to the mechanical load and $P_{loss} = T_{loss}\ \omega_{syn}$ is the mechanical power loss of the system.

Similar to the case of a generator, the electromagnetic power is the amount of power being converted from the electrical into the mechanical power.

i.e., $P_{em} = 3\ E_a\ I_a\ \cos\phi_{EaIa} = T\omega_{syn}$

Where $\phi E_a I_a$ is the angle between phasors E_a and I_a.

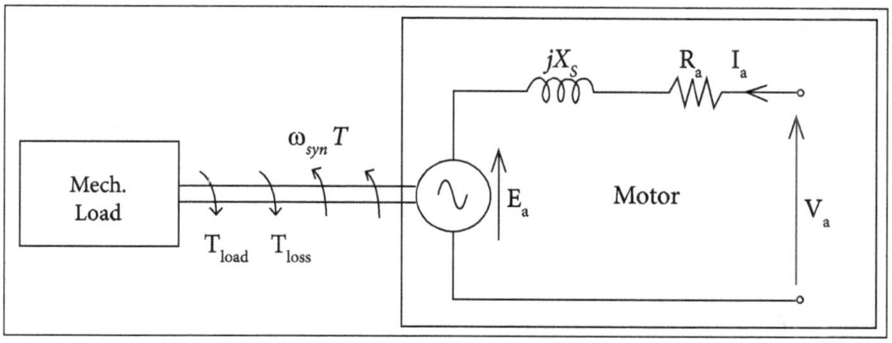

A synchronous machine operated as motor.

When the stator winding resistance is ignored, the per phase circuit equation can be approximately written as:

$$V_a = E_a + jX_s Ia$$

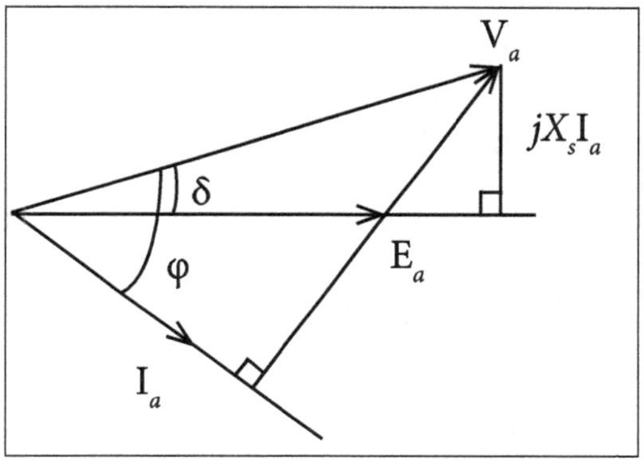

Motor phasor diagram.

From the phasor diagram, we can readily obtain:

$$V_a \sin\delta = X_s\, I_a \cos \phi_{EaIa}$$

Where,

$$\phi_{EaIa} = \phi - \delta$$

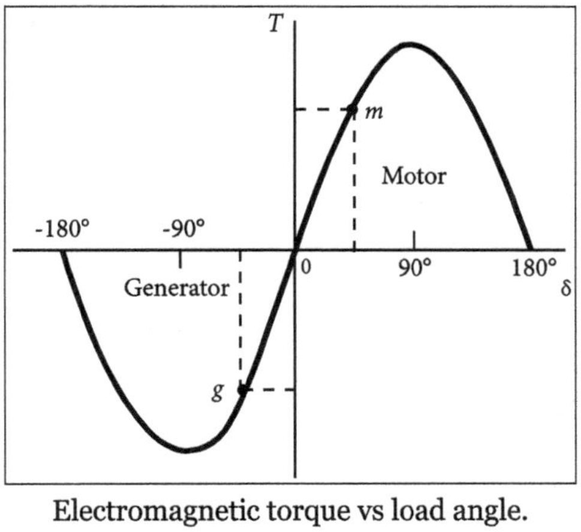

Electromagnetic torque vs load angle.

Therefore,

$$P_{em} = \frac{3E_a V_a}{X_s} \sin\delta$$

and

$$T = \frac{P_{em}}{\omega_{syn}} = \frac{3E_a V_a}{\omega_{syn} X_s} \sin\delta$$

Where δ is the load angle.

When the stator winding resistance is ignored, δ can be regarded as the angle between the rotor and stator rotating magnetic fields. In motor mode, the stator field is ahead of the rotor.

The electromagnetic torque of the synchronous machine is proportional to the sine function of the load angle, as plotted in the diagram above, where the curve in the third quadrant is for the situation when the machine is operated as a generator. The electromagnetic torque is negative because the armature current direction is reversed.

Phasor Diagram of Synchronous Motor

The characteristics of the synchronous generator on infinite bus bars are quite different from those when it is connected to another alternator and both are in parallel. When two alternators are connected in parallel, we have seen that a change in the excitation changes the terminal voltage and pf is determined by load.

Although, change in excitation for an alternator connected to infinite bus bar will not change the terminal voltage but the power factor is affected where the power developed by an alternator depends only on the mechanical power input. Now, consider the effect of excitation and driving torque on the performance of an alternator which is connected to infinite bus bar.

Effect of Excitation

Let us consider the cylindrical rotor as shown in the figure connected to the infinite busbar.

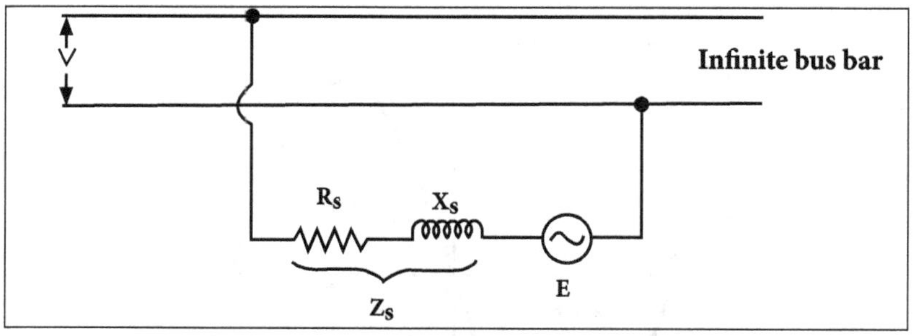

Cylindrical rotor connected to infinite bus bar.

The voltage equation can be written as:

$$\bar{E} = \bar{V} + \bar{I}\bar{Z}_s$$

Where,

E = Induced emf or excitation e.m.f.

I = Armature current.

V = Constant bus voltage.

Z_s = Synchronous impedance.

The same equation neglecting the armature resistance will be rewritten as:

$$\bar{E} = \bar{V} + j\bar{I}\bar{X}_s$$

Again we will consider the two cases, one with alternator on no load and other with alternator on load.

Again we will consider the two cases, one with alternator on no load and other with alternator on load.

Alternator on No Load

Since we are considering the losses to be zero, the power angle will be zero. Therefore, the power transferred from or to the bus is zero.

$$P = (EV/Xs)\sin\delta$$

Now, if the excitation is properly adjusted at no load, then the induced emf E will be equal to bus voltage V and no current will flow. This is the floating condition of alternator.

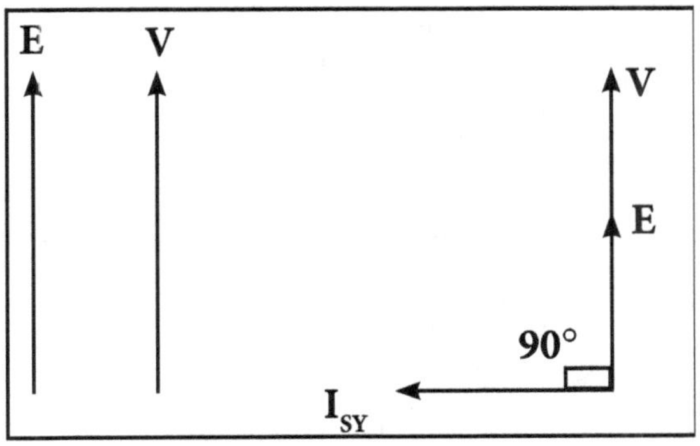

Excitation at no-load condition.

Now, if the alternator is under excitation, then induced emf E will be less than V. This will cause circulating current I_{SY} to flow which will lead E by angle of 90°. Due to this, it produces magnetizing mmf which will try to increase the field mmf to maintain the alternator terminal voltage equal to the bus bar voltage. This is shown in the above figure.

Similarly, if alternator is over excited, then the induced emf will be more than V which will again cause a circulating current I_{SY} to flow. The power angle δ is zero. This current lags E by 90°. This will produce demagnetizing armature mmf which will counterbalance the effect of increased field mmf and again the terminal voltage of an alternator can be equal to constant bus bar voltage V. This is represented in the figure.

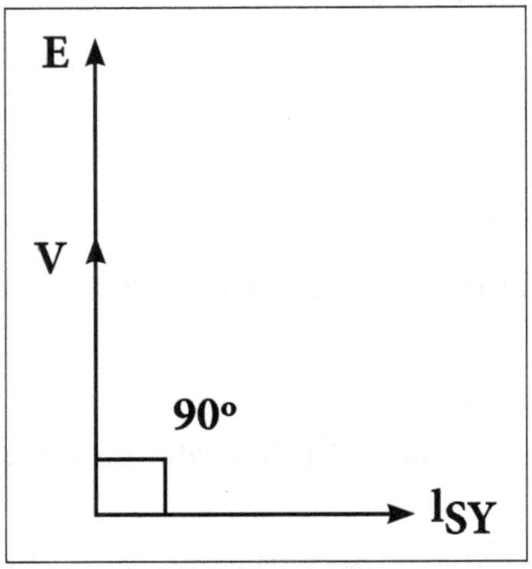

E Vs I_{SY} Characteristics.

It can be seen that in both the cases considered above, no active power is delivered since I_{SY} is in quadrature with V and load angle is also zero. But, alternator takes reactive power from bus since E < V and delivers it to bus if E > V.

Alternator on Load

Consider that alternator is supplying power to an infinite bus that has induced emf E, power angle δ and working at unity power factor with current I. With mechanical power input to the alternator remaining constant, the power given by $(EV/X_s) \sin \delta$ will remain constant. If by varying excitation induced emf E is increased to E_1, then the load angle will also change from δ to δ_1.

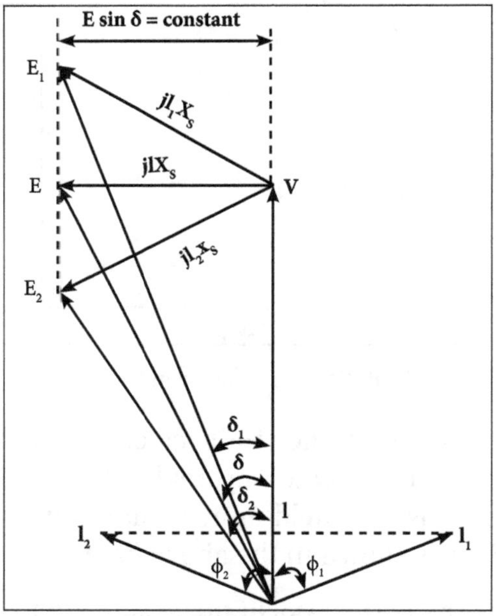

Phasor Diagram.

From the phasor diagram, it can be determined $E_1 \sin\delta_1 = E \sin\delta$ as V and X_s are constant. The drop due to synchronous reactance also increases and armature current increases from I to I_1. This current has two components such as one real component and other quadrature component. This quadrature component is nothing but demagnetizing component. This will result in lagging power factor $\cos\phi_1$.

Similarly if the excitation is decreased so that the induced emf reduces from E to E_2 with corresponding change in power angle from δ to δ_2. The armature current in this case will be I_2 which has real component and magnetizing component which results in leading power factor $\cos\phi_2$. This can be represented in the phasor diagram shown in figure.

From the phasor diagram it can be seen that:

$$I_1 \cos\phi_1 = I_2 \cos\phi_2 = 1$$

Multiplying by V throughout:

$$V I_1 \cos\phi_1 = V I_2 \cos\phi_2 = VI$$

This indicates that power delivered to the bus will remain constant. Therefore by changing the field excitation the active power is unaltered.

1.4 Dynamic Analysis of Synchronous Machine

Dynamic Analysis and Modeling of Synchronous Machine

The simplest model of a synchronous generator is a constant voltage source behind proper reactance. In this model, the voltage is assumed to be constant, only its phase angle changes. This model is very useful for the purpose of transient analysis involving short period i.e., one second or less. In this model, it is obvious that the changes in the flux linkages and saliency have been neglected. The machine equation becomes:

$$E_g' = V + I_a r_a + jI_a X' \qquad ...(1)$$

Where,

V is the full load terminal voltage.

E_g' is the transient voltage of the synchronous generator.

I_a is the full load armature current.

X_s' is the transient reactance.

The equivalent circuit and phasor diagram in this case will be as shown in the below figures (a) and (b) respectively. In order to include the effect of saliency, we calculate a fictitious voltage "E_q" located at the quadrature axis of the synchronous generator. Direct axis is taken along the main pole axis while quadrature axis lags the direct axis by 90°.

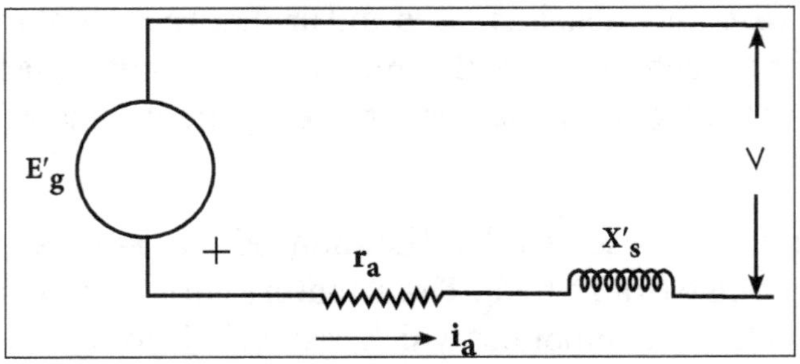

(a)

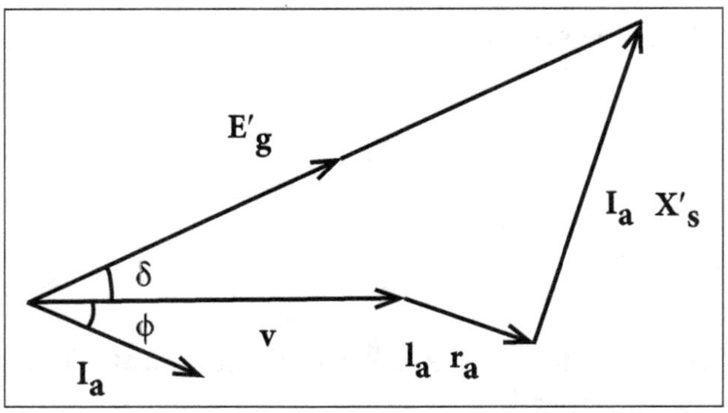

(b)

In the figure, δ is the torque angle which is assumed to change.

The expression for the voltage E_q in terms of full load terminal voltage V and full load armature current I_a will be:

$$E_q = V + I_a r_a + jI_a X_q \quad ...(2)$$

Here X_q is the quadrature axis synchronous reactance.

The equivalent circuit and the phasor diagram for this case are as shown in figure (c) and (d) respectively.

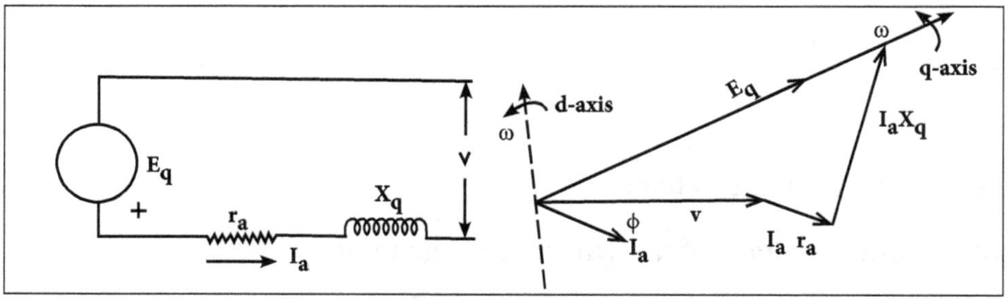

Equivalent circuit and phasor diagram.

In the figure, ω is rotor angular velocity which is equal to the synchronous speed ω_o only during normal operation.

The voltage E_g will be calculated as follows:

$$E_g = V + I_a r_a + jI_{ad} X_d + jI_{aq} X_q \quad ...(3)$$

Where,

X_q is the q-axis armature synchronous reactance i.e., $X_q = X_t + X_{aq}$.

X_d is the d-axis armature synchronous reactance i.e., $X_d = X_t + X_{ad}$.

X_{aq} is the q-axis armature magnetizing reactance corresponding to the q-axis component of the armature reaction flux i.e., ϕ_{aq}.

X_{ad} is the d-axis armature magnetizing reactance corresponding to the d-axis component of the armature reaction flux i.e., ϕ_{ad}.

Thus, the phasor diagram including the effect of saliency will be as shown in the below figure:

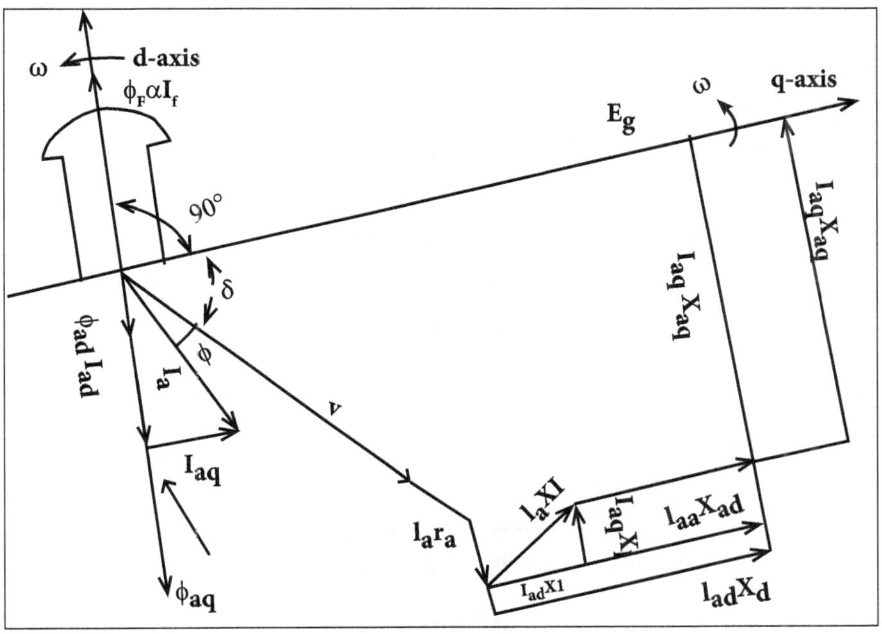

Phasor diagram.

The flux linkage of a circuit or a portion thereof, produced by the current in the same circuit element, is considered to have the same sign as the current. Voltage induced by the flux linkage is given by:

$$V = \frac{d\psi(t)}{dt} \qquad ...(4)$$

Here $\psi(t)$ is the flux linkage and it is equal to Real $\psi_{max}e^{j\omega t}$.

Then the voltage will be:

$$V(t) = \text{Re}\left(-j\omega\psi_{max}e^{j\omega t}\right) \qquad ...(5)$$

i.e., The voltage phasor lags the flux phasor by 90°.

Circuit Equations

Each winding k in the machine can be described as,

$$V_k(t) = -R_k i_k(t) - \frac{d\ k}{dt} \qquad \text{...(6)}$$

Where,

V_k is the voltage across the terminal of the winding.

i_k is the current out of rotor and stator i.e., armature winding.

Some assumptions are usually made to develop the detailed model of the synchronous machines. They are as follows:

- All Inductances are independent of current.
- The air-gap shape and distribution of windings are such that all self and mutual inductances may be represented as constants plus sinusoidal functions of the electrical rotor positions.
- Higher order time and space harmonics are neglected.

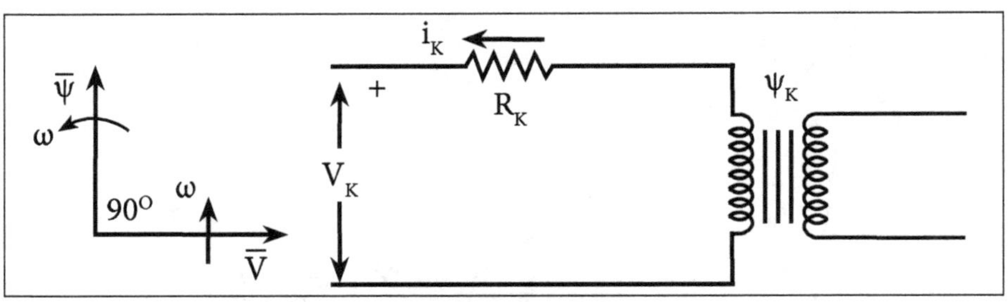

The State Space Synchronous Machine Model.

To describe the dynamic behavior of synchronous machines in time domain, let us consider the state space modeling concept. Therefore, based on the dynamic equations of synchronous machines and their particular state variables, the state space model can be utilized to determine the future state of the machine provided that the present state and the excitation signals are known.

Initially, consider a general non-linear system with multiple states and inputs as:

$$\dot{x} = f(x_1, x_2, \ldots.., x_n, u_1, u_2, \ldots, u_m) \qquad \text{...(1)}$$

Where,

x_i is the ith vector of the state variables.

u_j is the jth system driving variable.

f is a set of non-linear functions.

Suppose the equilibrium points of x_{oi} and u_{oj} are defined such that $f(x_{o1},\ldots, x_{on}, \ldots, u_{o1}, \ldots, u_{om}) = 0$. If x_i and u_j are considered to be a perturbed state of the above system:

$$\left.\begin{array}{llll} x_1 = x_{o1} + \tilde{x}_1 & x_2 = x_{o2} + \tilde{x}_2 & \cdots & x_\eta = x_{o\eta} + \tilde{x}_\eta \\ u_1 = u_{o1} + \tilde{u}_1 & u_2 = u_{o2} + \tilde{u}_2 & \cdots & u_m = u_{om} + \tilde{u}_m \end{array}\right\} \qquad \ldots(2)$$

Note that at the equilibrium points, the function f is zero. The linearization of the system about the equilibrium point can be obtained using Taylor series expansion and by ignoring the second and higher order terms as:

$$\dot{x} = f(x_1,\ldots x_n, u_1,\ldots,u_m) \cong \sum_{i=1}^{n} \frac{\partial f}{\partial x_i}\bigg|_{\left(\substack{x_i = x_{oi} \\ u_i = u_{oi}}\right)} \tilde{x}_i + \sum_{i=1}^{n} \frac{\partial f}{\partial u_i}\bigg|_{\left(\substack{x_i = x_{oi} \\ u_i = u_{oi}}\right)} \tilde{u}_i \qquad \ldots(3)$$

Thus, for small perturbation of the non-linear system around the equilibrium point, the linear state space model of the system can be written as:

$$\begin{array}{l} \dot{X} = AX + BU \\ Y = CX + DU \end{array} \qquad \ldots(4)$$

In which A, B, C and D are termed as the system, input, output and feed-forward coefficient matrices, respectively.

Considering X_o as the initial condition of the system, applying the Laplace transformation to the state space equations in (3), we have:

$$sX(s) - X_o = AX(s) + BU(s) \qquad \ldots(5)$$

Therefore,

$$(sI - A)X(s) = X_o + BU(s) \qquad \ldots(6)$$

$$X(s) = (sI - A)^{-1} X_o + (sI - A)^{-1} BU(s) \qquad \ldots(7)$$

It can be proven that:

$$(sI - A)^{-1} = L(e^{At}) \qquad \ldots(8)$$

Therefore, state space equations in time-domain can be obtained as:

$$\left.\begin{array}{l} X(t) = e^{A(t-t_o)} X_o + \int_{t_o}^{t} e^{A(t-\tau)} BU(\tau)d\tau \\ Y(t) = CX(t) + DU(t) \end{array}\right\} \qquad \ldots(9)$$

1.4.1 Dynamic Performance of Synchronous Machine

Dynamic Performance During a Sudden Change in the Input Torque

It is instructive to observe the dynamic performance of the synchronous machine during a step change in its input torque. For this purpose, the differential equation describes that the synchronous machine were programmed on a computer and a study was performed.

Two large machines are considered. One is a low-speed hydro turbine generator and a high-speed steam turbine generator. In case of hydro turbine generator, parameters are given for only one damper winding in the q axis.

Hydro Turbine Generator

Line-to-line voltage: 20kV

Rating: 325 MVA

Poles: 64

Power factor: 0.85

Speed: 112.5r/min

Combined inertia of generator and turbine: $J = 35.1 \times 10^6 \, J.s^2$

Or $WR^2 = 833.1 \times 10^6$ Ibm.ft^2

$H = 7.5s$

Parameters in Ohms Per Unit

$X_{ls} = 0.1478\Omega$, 0.120pu

$r_s = 0.00234\Omega$, 0.0019pu

$X_q = 0.5911\Omega$, 0.480pu

$X_d = 1.0467\Omega$, 0.850pu

$X'_{lfd} = 0.2523\Omega$, 0.2049pu

$r'_{fd} = 0.00050\Omega$, 0.00041pu

$r'_{kd} = 0.01736\Omega$, 0.0141pu

$r'_{kq2} = 0.01675\Omega$, 0.0136pu

$X'_{lkd} = 0.1970\Omega$, 0.160pu

$X'_{lkq2} = 0.1267\Omega$, 0.1029pu

The computer traces shown in figure (a) and (b) illustrates the dynamic behavior of the hydro-turbine generator following a step change in its input torque from zero to 27.6 x 10^6N.m.

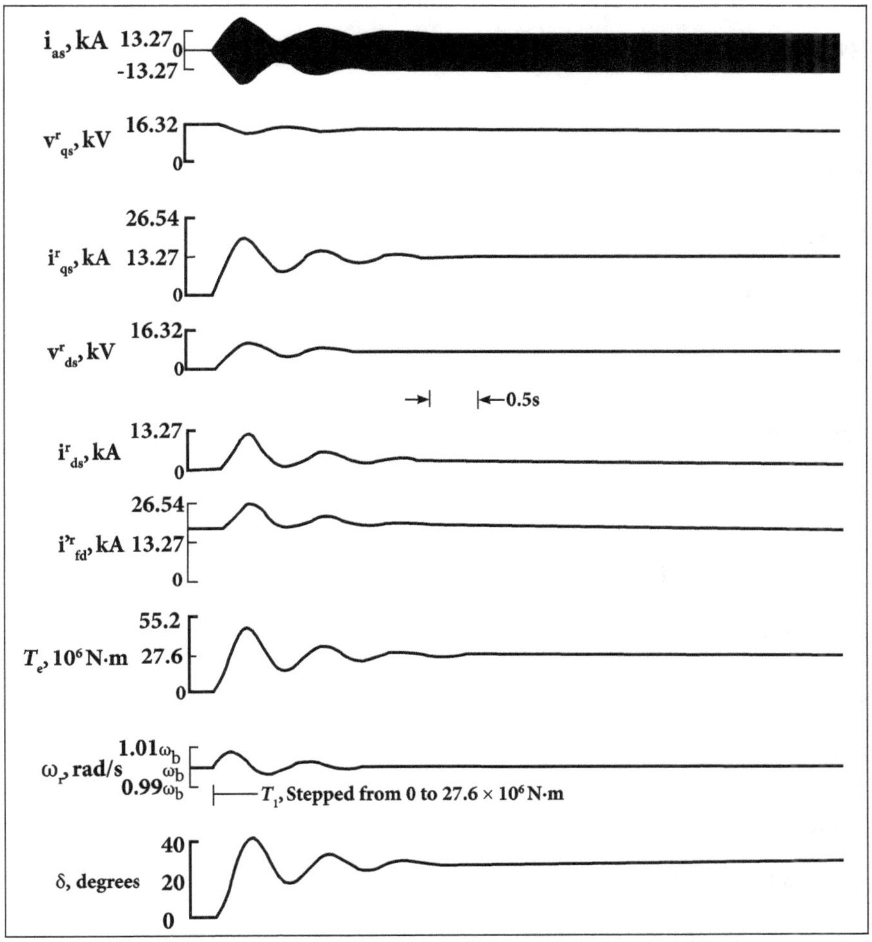

(a) Dynamic performance of a hydro turbine generator during a step increase in input torque from zero to rated.

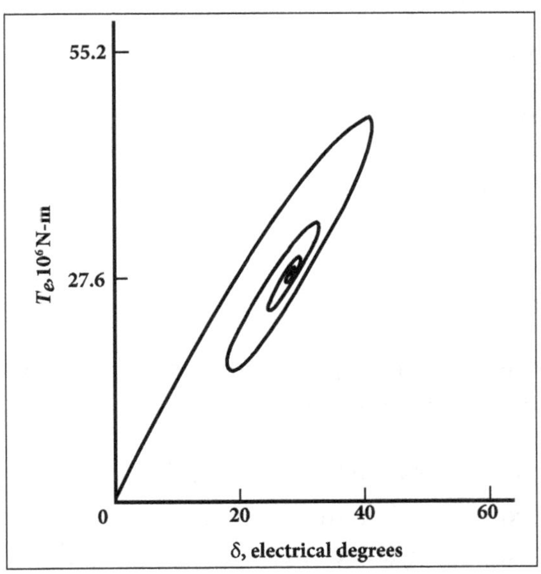

(b) Torque versus rotor angle characteristics.

Steam Turbine Generator

Line-to-line voltage: 26 kV

Rating: 835 MVA

Poles: 2

Power factor: 0.85

Speed: 3600 r/min

Combined inertia of generator and turbine:

$J = 0.0658 \times 10^6 \text{ J.s}^2$

Or $WR^2 = 1.56 \times 106 \text{ lbm.ft}^2$

$H = 5.6\text{s}$

Parameters in Ohms and Per Unit

$r_s = 0.00243\Omega$, 0.003pu

$X_{ls} = 0.1538\Omega$, 0.19 pu

$X_d = 1.457\Omega$, 1.8 pu

$X_q = 1.457\Omega$, 1.8 pu

$X'_{lkq1} = 0.6578\Omega$, 0.8125 pu

$X'_{lfd} = 0.1145\Omega$, 0.1414 pu

$r'_{kq1} = 0.00144\Omega$, 0.00178 pu

$r'_{fd} = 0.00075\Omega$, 0.000929 pu

$X'_{lkq2} = 0.07602\Omega$, 0.0939 pu

$X'_{lkd} = 0.06577\Omega$, 0.08125 pu

$r'_{kq2} = 0.00681\Omega$, 0.00841 pu

$r'_{kd} = 0.01080\Omega$, 0.01334 pu

The dynamic behavior of the steam turbine generator is depicted in figure (c) and (d). In this case, the step change in input torque is from zero to 1.11 x 10^6 N.m.

In figure (a) and (c), the following variables are plotted: i_{as}, $v^r_{qs}, i^r_{qs}, v^r_{ds}, i^r_{ds}, i^{Ir}_{fd}, T_e, \omega_r$ and δ where ω_r is in electrical radians per second and δ is in electrical degrees. Figures (b) and (d) illustrate the dynamic torque versus rotor angle characteristics. In all figures, the scales of the voltages and currents are given in multiples of peak rated values.

In both the study, it is assumed that the machine is connected to a bus whose voltage and frequency remain constant, at the rated values, regardless of the stator current. This is commonly referred to as an infinite bus, because its characteristics do not change regardless of the power supplied or consumed by any device connected to it.

Although an infinite bus cannot be realized in practice, its characteristics are approached if the power delivery capability of the system, at the point where the machine is connected, is much larger than the rating of the machine. Initially, each machine is operating with zero input torque with the excitation held fixed at the value that gives rated open-circuit terminal voltage at synchronous speed.

It is instructive to observe the plots of T_e, ω_r and δ following the step change input torque. In particular, consider the response of the hydro turbine generator where the machine is subjected to a step increase in input torque from zero to 27.6 x 10^6 N.m.

The rotor speed begins to increase immediately following the step increase in input torque, whereas the rotor angle increases. The rotor speeds up until the accelerating torque on the rotor is zero. As noted in figure(a), the speed increases to approximately 380 electrical radians per second, at which time T_e is equal to T_I because the change of ω_r is zero and hence the inertial torque (T_{IT}) is zero.

Even though the accelerating torque is zero at this time, the rotor is running above synchronous speed. Hence δ and thus T_e, will continue to increase. The increase in T_e, which is an increase in the power output of the machine, causes the rotor to decelerate towards its synchronous speed. However, when synchronous speed is reached, the magnitude of δ has become larger than necessary to satisfy the input torque.

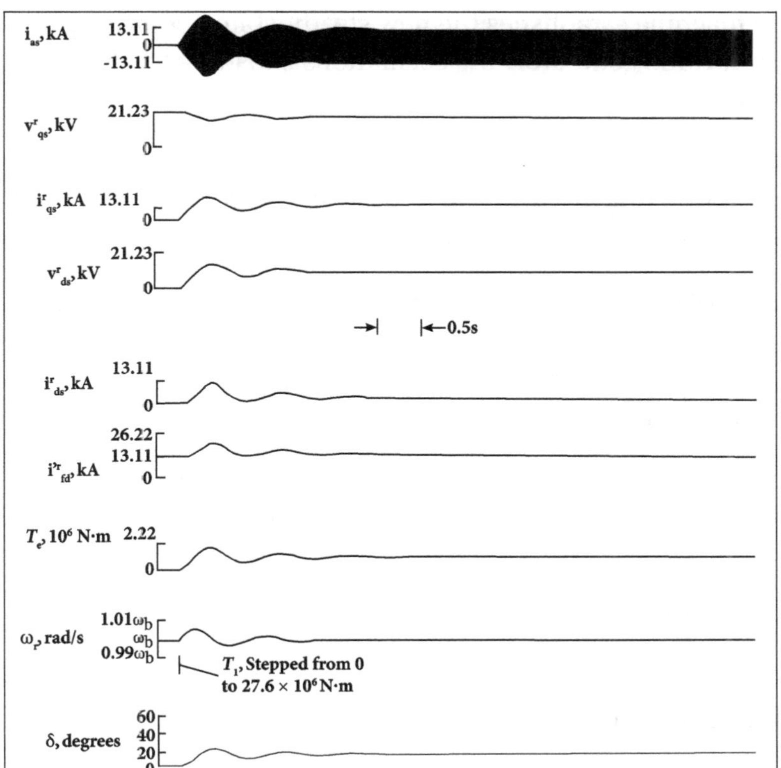

(c) Dynamic performance of a steam turbine generator during a step increase in input torque from zero to 50% rated.

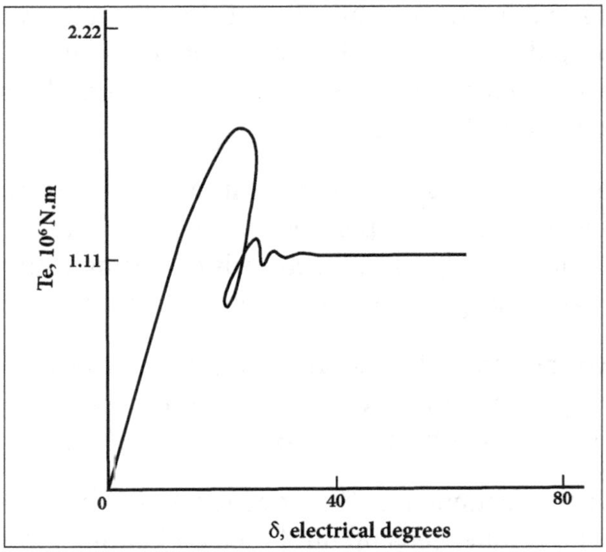

(d) Torque versus rotor angle characteristics.

Note that at the first synchronous speed crossing of ω_r after the change in input torque, δ is approximately 42 electrical degrees and T_r is approximately 47 x 10^6 N.m. Hence, the rotor continues to decelerate below synchronous speed and consequently δ begins to decrease, which in turn decreases T_e. Damped oscillations of the machine variables continue and a new steady-state operating point is finally attained.

In the case of the hydro turbine generator, the oscillations in machine variables subside in a matter of 2 or 3s and the machine establishes the new steady-state operating point within 8 or 10s. In the case of the steam turbine generator, the oscillations subside rapidly, but the new steady-state operating point is slowly approached.

The damping is a function of the damper windings and can be determined from an eigenvalue analysis. This rather slowly approach to the new steady-state operating point in the case of the steam turbine generator is also apparent from the plot of T_e versus δ.

For a moment, the expression for steady-state torque:

$$T_e = \left(\frac{3}{2}\right)\left(\frac{P}{2}\right)\left(\frac{1}{\omega_b}\right)\left[\frac{E_{xfd}^{Ir}\sqrt{2}\,V_s}{(\omega_e/\omega_b)X_d}\sin\delta + \left(\frac{1}{2}\right)\left(\frac{\omega_e}{\omega_b}\right)^{-2}\left(\frac{1}{X_q}-\frac{1}{X_d}\right)\left(\sqrt{2}V_s\right)^2\sin 2\delta\right]$$

For the hydro turbine generator with $E_{xfd}^{Ir} = \sqrt{\frac{2}{3}}20\,\text{kV}$, we have:

$$T_e = (32.5\sin\delta + 12.5\sin 2\delta)\times 10^6\ \text{N.m} \qquad \text{...(1)}$$

For the steam turbine generator with $E_{xfd}^{Ir} = \sqrt{\frac{2}{3}}26\,\text{kV}$, we obtain:

$$T_e = 1.23 \times 10^6 \sin\delta\ \text{N.m} \qquad \text{...(2)}$$

If (1) and (2) are plotted on figures (b) and (d), respectively, the steady-state T_e versus δ curves will pass through the final value of the dynamic T_e versus δ plots. However, the dynamic torque-angle characteristics immediately following the input torque disturbance yields a much larger T_e for a given value of δ than does the steady-state characteristic.

In other words, the dynamic or transient torque-angle characteristics are considerably different from the steady-state characteristic and the steady-state T_e versus δ curve applies only after all transients have subsided. Although the computation of the transient torque during speed variations requires the solution of nonlinear differential equations, it can be approximated quite simply.

Motor action, wherein a load torque is applied to the machine, would essentially yield the mirror image of the Te versus δ plots differing only by the ohmic losses.

1.4.2 Three - phase Fault

Fault

A fault in a circuit is any failure which interferes with the normal flow of current. The faults are associated with abnormal change in current, voltage and frequency of the power system.

Faults Occur in a Power System

The faults occur in a power system due to:

- Insulation failure of equipment.
- Flashover of lines initiated by a lighting stroke.
- Due to permanent damage to conductors and towers or due to accidental faulty operations.

Various Types of Faults

- Series fault or open circuit fault.
  - One open conductor fault.
  - Two open conductor fault.
- Shunt fault or short circuit fault.
- Symmetrical fault or balanced fault.
  - Three phase fault.
- Unsymmetrical fault or unbalanced fault.
  - Line to ground (L-G) fault.
  - Line to Line (L-L) fault.
  - Double line to ground (L-L-G) fault.

Types of fault	Relative frequency of occurrence of faults
Three phase fault	5%
Line to ground (L-G) fault	10%
Line to Line (L-L) fault	15%
Double line to ground (L-L-G) fault	70%

Symmetrical Fault or Balanced Three Phase Fault

This type of fault is defined as the simultaneous short circuit across all the three phases. It occurs infrequently, but it is the most severe type of fault encountered. Because the network is balanced, it is solved by per phase basis using Thevenins theorem or bus impedance matrix or KVL, KCL laws.

Abnormal System Behavior under Conditions of Symmetrical Short Circuit

Such conditions are caused in the system accidentally through the insulation failure of equipment or flashover of lines initiated by the lightning stroke or through accidental faulty operation. The system must be protected against the flow of heavy short circuit currents by disconnecting the faulty part of the system by means of circuit breakers operated by protective relaying.

For proper choice of circuit breakers and protective relaying, we must estimate the magnitude of currents which may flow under the short circuit condition. This is the scope of fault analysis. The majority of system faults is not three-phase faults, but faults involving one line to ground or occasionally two lines to ground.

Though the symmetrical faults are rare, the symmetrical fault analysis must be carried out, as this type of fault generally leads to most severe fault current flow against which the system must be protected. Symmetrical fault analysis is simpler to carry out.

A power network comprises synchronous generators, transformers, lines and loads. Though the operating conditions are important, the loads can be neglected during fault, as voltages dip very low so that currents drawn by loads can be neglected in comparison to fault currents. The synchronous generator during short circuit has a characteristic time-varying behavior.

In the event of a short circuit, the flux per pole undergoes dynamic change with associated transients in damper and field windings. The reactance of the circuit model of the machine changes in the first few cycles from a low sub-transient reactance to a higher transient value. Finally, settling at a still higher synchronous value.

Depending upon the arc interruption time of circuit breakers, a suitable reactance value is used for the circuit model of synchronous generators for short circuit analysis. In a modern large interconnected power system, heavy currents flowing during a fault must be interrupted much before the steady state conditions are established.

Furthermore, from the considerations of mechanical forces that act on circuit breaker components, the maximum current that a breaker has to carry momentarily must also be determined. For selecting a circuit breaker we must determine the initial current that flows on occurrence of a short circuit and also the current in the transient that flows at the time of circuit interruption.

Basic Assumptions in Fault Analysis of Power Systems

- Representing of each machine by a constant voltage source behind the proper reactance which may be X", X' or X.
- Pre-fault load current are neglected.
- Transformer taps are assumed to be nominal.
- Shunt elements in the transformers model that account for magnetizing current and core loss are neglected.
- A symmetric three phase power system is conducted.
- Shunt capacitance and series resistance in transmission are neglected.
- The negative sequence impedances of alternators are assumed to be the same as their positive sequence impedance Z+ = Z-.

Fault Analysis of 3φ Generator

Consider a 3φ generator with a fault through impedance Z_f between phases b and c shown. Assume the generator is unloaded (no load). The conditions at fault bus k are expressed by following relationship:

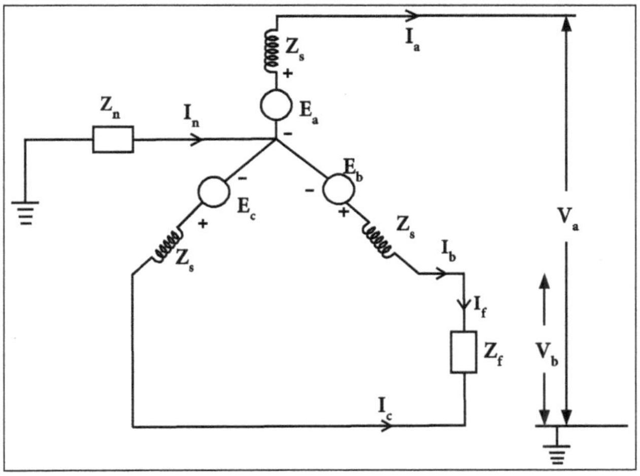

$$I_b = -Ic$$

$I_a = 0$, unloaded generator.

$$V_b - V_c = Z_f I_b \Rightarrow V_c = V_b - Z_f Ib$$

Substitute for $I_b = I_c$, $I_a = 0$, the symmetrical components of currents are:

$$\begin{bmatrix} I_a^0 \\ I_a^+ \\ I_a^- \end{bmatrix} = \frac{1}{3}\begin{bmatrix} 1 & 1 & 1 \\ 1 & a & a^2 \\ 1 & a^2 & a \end{bmatrix}\begin{bmatrix} I_a \\ I_b \\ I_c \end{bmatrix}$$

$$\begin{bmatrix} I_a^0 \\ I_a^+ \\ I_a^- \end{bmatrix} = \frac{1}{3}\begin{bmatrix} 1 & 1 & 1 \\ 1 & a & a^2 \\ 1 & a^2 & a \end{bmatrix}\begin{bmatrix} 0 \\ +I_b \\ -I_b \end{bmatrix}$$

$$I_a^0 = 1/3[0 + I_b - I_b] = 0$$

$$I_a^+ = 1/3\left[aI_b - a^2 I_b\right]$$

$$I_a^- = 1/3\left[a^2 I_b - aI_b\right]$$

$$I_a^+ = -I_a^- \;\&\; I_a^0 = 0$$

From the sequence networks of the generator, the symmetrical voltages are given by:

$$\begin{bmatrix} V_a^0 \\ V_a^+ \\ V_a^- \end{bmatrix} = \begin{bmatrix} 0 \\ E_a \\ 0 \end{bmatrix} - \begin{bmatrix} Z_{kk}^0 & 0 & 0 \\ 0 & Z_{kk}^+ & 0 \\ 0 & 0 & Z_{kk}^- \end{bmatrix}\begin{bmatrix} I_a^0 \\ I_a^+ \\ I_a^- \end{bmatrix}$$

$$V_a^0 = -Z_{kk}^0 \; I_a^0 = 0$$
$$V_a^+ = E_a \; Z_{kk}^+ \; I_a^+$$
$$V_a^- = + Z_{kk} \; I_a^+$$

The phase currents are given by:

$$\begin{bmatrix} I_a \\ I_b \\ I_c \end{bmatrix} = \begin{bmatrix} 1 & 1 & 1 \\ 1 & a^2 & a \\ 1 & a & a^2 \end{bmatrix}\begin{bmatrix} I_a^0 \\ I_a^+ \\ I_a^- \end{bmatrix}$$

$$= \begin{bmatrix} 1 & 1 & 1 \\ 1 & a^2 & a \\ 1 & a & a^2 \end{bmatrix}\begin{bmatrix} 0 \\ I_a^+ \\ -I_a^+ \end{bmatrix}$$

$$I_a = 0 \; I_b = a^2 \; I_a^+ - aI_a^+ = I_a^+\left(a^2 - a\right)$$

$$I_a = aI_a^+ - a^2 \; I_a^T = I_a^+\left(a - a^2\right)$$

The voltage throughout the zero sequence network must be zero, since there are no zero sequence sources and because $I_a^o = 0$ current is not being injected into that network due to the fault. Hence L-L fault calculation does not involve zero sequence network.

The phase voltages are:

$$\begin{bmatrix} V_a \\ V_b \\ V_c \end{bmatrix} = \begin{bmatrix} 1 & 1 & 1 \\ 1 & a^2 & a \\ 1 & a & a^2 \end{bmatrix} \begin{bmatrix} 0 \\ V_a^+ \\ V_a^- \end{bmatrix}$$

$$V_a^o = 0$$

$$V_a = V_a^+ + V_a^-$$

$$V_b = a^2 V_a + V_a^-$$

$$V_c = aV_a^+ + a^2 V_a^-$$

From the condition, $V_b - V_c = Z_f I_b$

Substitute V_b and V_c values:

$$a^2 V_a^+ + a V_a^- - a V_a^+ - a^2 V_a^- = Z_f I_b$$

$$V_a^+ + (a^2 - a) - V_a^- [a^2 - a] = Z_f I_b$$

$$(a^2 - a)[V_a^+ - V_a^-] = Z_f I_b$$

$$(a^2 - a)[V_a^+ - V_a^-] = (a^2 - a) I_a^+ Z_f$$

$$V_a^+ - V_a^- = I_a^+ Z_f$$

Substitute V_a^+, V_a^- values:

$$E_a - Z_{kk}^+ I_a^- - [-Z_{kk}^+ I_a^-] = I_a^+ Z_f$$

$$E_a - (Z_{kk}^+ + Z_{kk}^-) I_a^+ = I_a^+ Z_f$$

$$E_a = I_a^+ = I_a^+ [Z_f + Z_{kk}^+ + Z_{kk}^-]$$

$$X'_d = 0.2591 \text{ ohms}$$

$$I_a^- = -I_a^+$$

$$I_a^o = 0$$

Current in phase domain is:

$$I_b = -I_c = \left(a^2 - a\right) I_a^+$$

$$= \left(-0.5 - j0.866 + 0.5 - j0.866\right) I^4$$

$$= j\sqrt{3}\, I_a^+$$

$$\therefore\ I_f = \frac{-j\sqrt{3}E_a}{Z_{kk}^+ + Z_{kk}^- + Z_f}$$

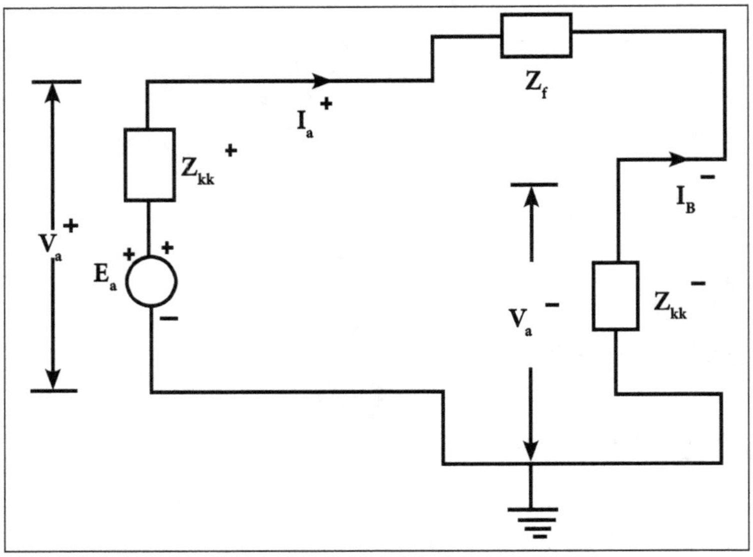

From the network:

$$I_a^- = -I_a^+,\ I_a^o = 0$$

$$I_a^+ = \frac{E_a}{Z_{kk}^+ + Z_{kk}^- + Z_f}$$

$$V_a^+ = V_a^- + Z_f I_a^+$$

$$I_f = I_b = I_a^+ \left(a^2 - a\right) = -i,\sqrt{3}\ I_a^+$$

Direct short circuit L-L fault:

$$Z_f = 0,\ I_a = 0 \text{ and } I_b = -I_c$$

$$V_b = V_c$$

$$I_a^+ = -I_a^-$$

$$If = \frac{-j\sqrt{3}E_a}{Z_{kk}^{+} + Z_{kk}^{-}}$$

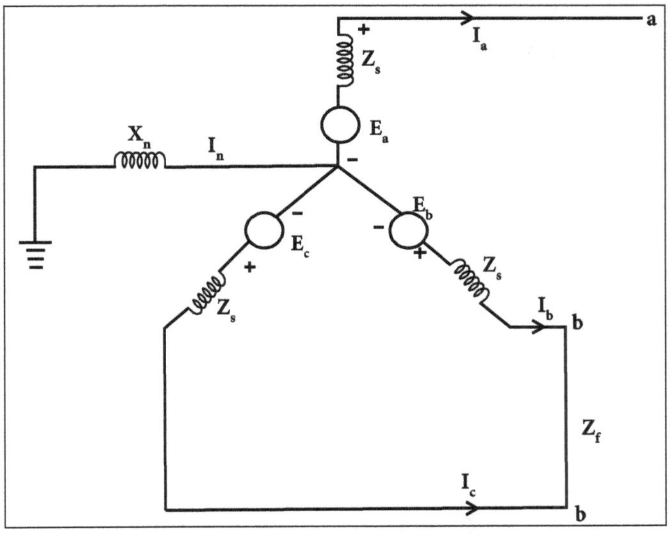

Transients on a Transmission Line

Consider a transmission line of resistance R and inductance L supplied by an AC source of voltage v, such that $v = V_m \sin(\omega t + \alpha)$ as shown in figure (a). Consider the short circuit transient on this transmission line.

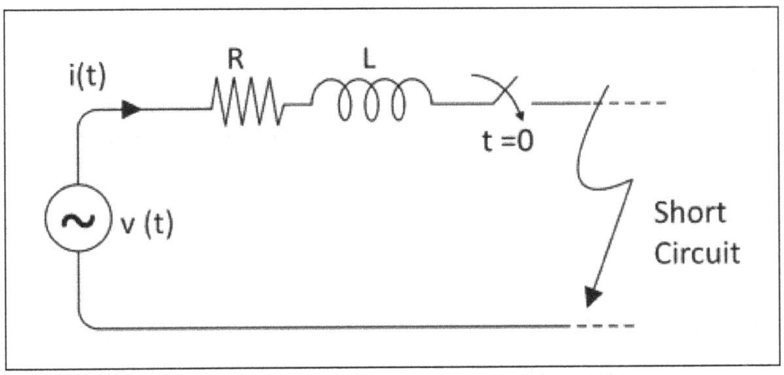

(a) Short Circuit Transients on an Unloaded Line.

To analyze this symmetrical 3-phase fault, the following assumptions are made:

- The short circuit occurs when the line is unloaded.
- The supply is a constant voltage source.
- The line capacitance is negligible.

Thus, the line can be modeled by a lumped R-L series circuit.

Let the short circuit take place at t = 0. The parameter, α controls the instant of short circuit on the voltage wave.

From the basic circuit theory, it is observed that the current after short circuit is composed of two parts such as $i = i_s + i_t$.

Where,

i_s is the steady state current.

I_t is the transient current.

Consider:

$$v = V_m \sin(\omega t + \alpha)$$

$$= iR + L\,(di/dt) \quad ...(1)$$

And

$$i = I_m \sin(\omega t + \alpha - \theta) \quad ...(2)$$

$$\text{Where } V_m = \sqrt{2}V;\ I_m = \sqrt{2}I;\ Z_{mag} = \sqrt{\left[R^2 + (\omega L)^2\right]} = \tan^{-1}(\omega L / R) \quad ...(3)$$

Thus, $$i_s = [V_m/Z]\ \sin(\omega t + \alpha - \theta) \quad ...(4)$$

Consider the performance equation of the circuit of figure (a) under circuit as:

$$iR + L\,(di/dt) = 0$$

i.e., $$(R/L + d/dt)i = 0 \quad ...(5)$$

To solve equation (5), consider the complementary part of the solution as:

$$CF = C^1 e^{(-t/\tau)} \quad ...(6)$$

Where,

τ (= L/R) is the time constant.

C_1 is a constant given by the value of steady state current at t = 0.

Thus, we have:

$$C_1 = -i_s\ (0)$$

$$= -[V_m/Z]\sin(\alpha - \theta)$$

$$= [V_m/Z]\sin(\theta - \alpha) \quad ...(7)$$

Similarly, the expression for the transient part is given by:

$$i_t = -i_s(0)\ e^{(-t/\tau)}$$

$$= [V_m/Z]\sin(\theta-\alpha)^{e(-R/L)t} \quad ...(8)$$

Thus, the total current under short circuit is given by the solution of equation (1) as,

$$i = i_s + i_t$$

$$= [\sqrt{2}V/Z]\sin(\omega t + \alpha - \theta) + [\sqrt{2}V/Z]\sin(\theta - \alpha)e^{(-R/L)t} \quad ...(9)$$

Thus, i_s is the sinusoidal steady state current termed as the symmetrical short circuit current and i_t is the unidirectional value which is termed as the DC off-set current. This causes the total current to be unsymmetrical till the transient decays, as shown in figure (b):

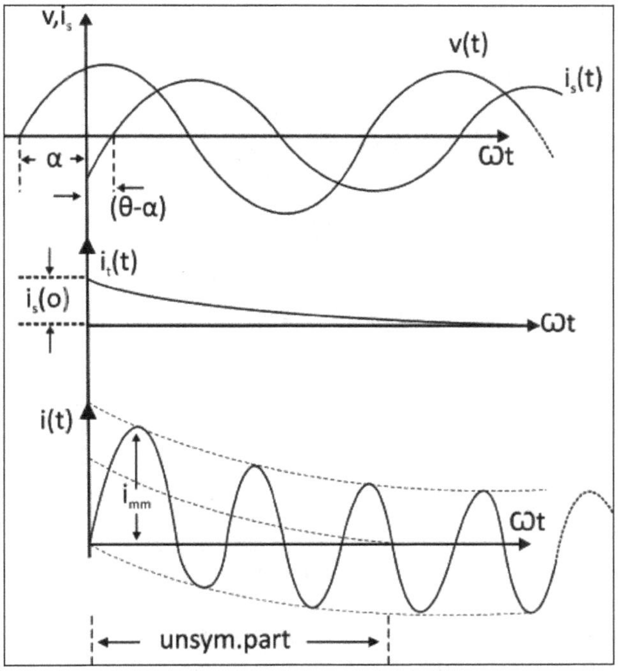

(b) Plot of Symmetrical short circuit current, i(t).

The maximum momentary current, i_{mm} thus corresponds to the first peak. Hence, if the decay in the transient current during this short interval of time is neglected, then we have:

$$i_{mm} = [\sqrt{2}V/Z]\sin(\theta - \alpha) + [\sqrt{2}V/Z] \quad ...(10)$$

Since the resistance of the transmission line is very small, the impedance angle θ, can be taken to be approximately equal to 90°. Hence, we have:

$$i_{mm} = [\sqrt{2}V/Z]\cos\alpha + [\sqrt{2}V/Z] \quad ...(11)$$

This value is maximum when the value of α is zero. This value corresponds to the short circuiting instant of the voltage wave when it is passing through zero. Thus, the final expression for the maximum momentary current is obtained as:

$$i_{mm} = 2[\sqrt{2}V/Z] \quad ...(12)$$

Hence, it is observed that the maximum momentary current is twice the maximum value of symmetrical short circuit current. This is referred as the doubling effect of the short circuit current during the symmetrical fault on the transmission line.

Dynamic Performance During a 3-phase Fault at the Machine Terminals

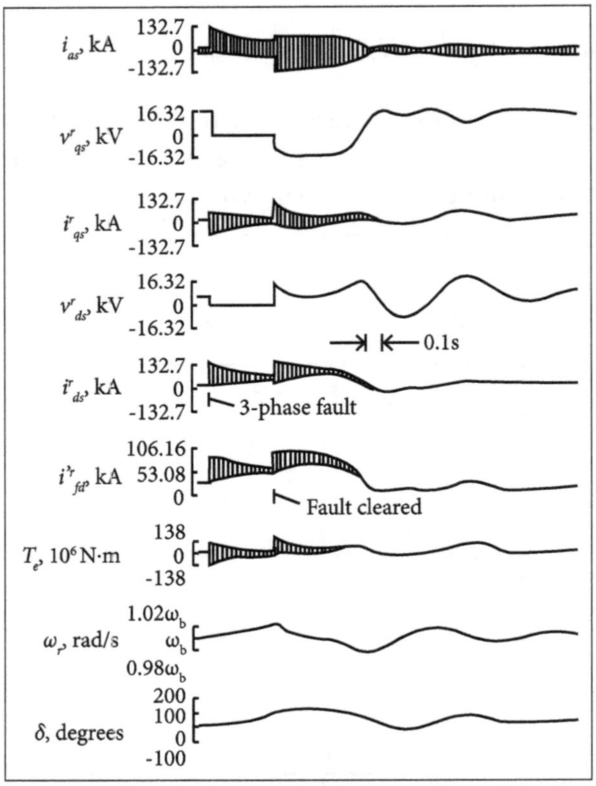

(a) Dynamic performance of a hydro turbine generator during a 3-phase fault at the terminals.

The stability of synchronous machines throughout a power system following a fault is of major concern. A 3-phase fault or short-circuit rarely occurs, and a 3-phase fault at the machine terminals is even more uncommon. Nevertheless, it is instructive to observe the dynamic performance of the synchronous machine during this type of a fault.

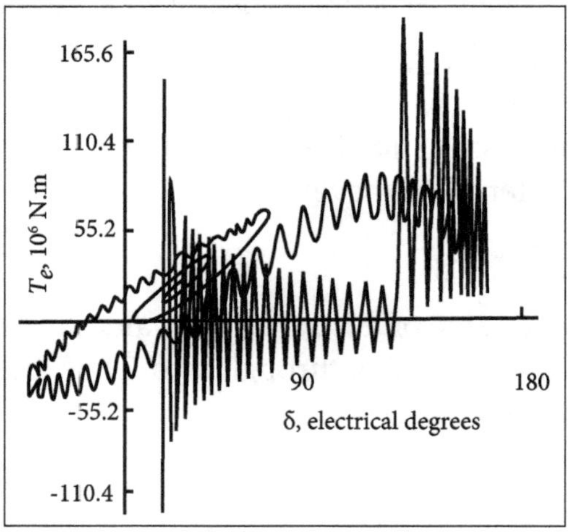

(b) Torque versus rotor angle characteristics.

The computer traces shown in figures (a) and (b) illustrate the dynamic behavior of the hydro turbine generator during and following a 3-phase fault at the terminals. The dynamic behavior of the steam turbine generator as a result of a 3-phase terminal fault is shown in figures (c) and (d).

In figure (a) and (c), the following variables are plotted:

$$i_{as}, v^r_{qs}, i^r_{qs}, v^r_{ds}, i^r_{ds}, i^{\prime r}_{fd}, T_e, \omega_r, \text{and}\, \delta$$

Figures (b) and (d) illustrate the dynamic torque-angle characteristics during and following the 3-phase fault. In each case, the machine is initially connected to an infinite bus delivering rated MVA at rated power factor. In the case of the hydro turbine generator, the input torque is held constant at (0.85)27.6 x 10⁶ N.m with $E^{\prime r}_{xfd}$ fixed at $(1.6)\sqrt{\frac{2}{3}}20\,\text{kV}$;

For the steam turbine generator, $T_I = (0.85)2.22 \times 10^6$ N.m and $E^{\prime r}_{xfd} = (2.48)\sqrt{\frac{2}{3}}26\,\text{kV}$.

With the machines operating in this steady-state condition, a 3-phase terminal fault is simulated by setting v_{as}, v_{bs} and v_{cs} to zero. In the simulation, at the instant v_{as} passes through zero going positive. The transient offset in the phase currents is reflected into the rotor reference-frame variables and the instantaneous torque as a decaying 60-Hz pulsation.

Because the terminal voltage is zero during the 3-phase fault, the machine is unable to transmit power to the system. Hence, all of the input torque, with the exception of the ohmic losses, accelerates the rotor.

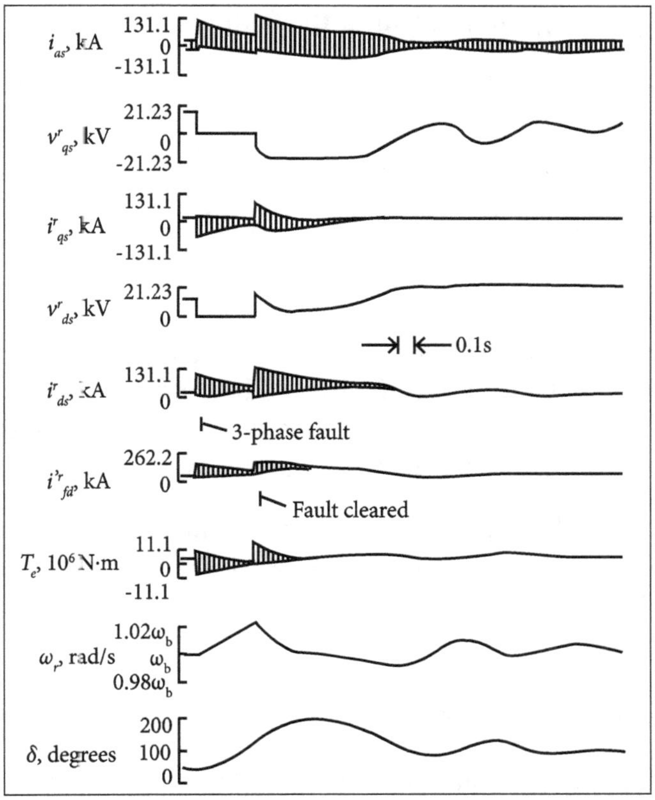

(c) Dynamic performance of a steam turbine generator during a 3-phase fault at the terminals.

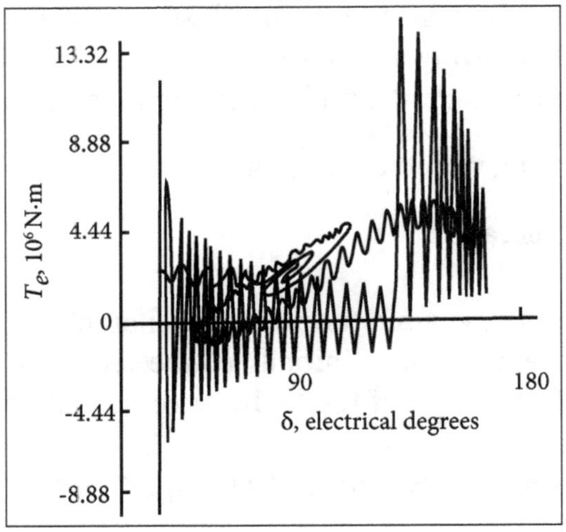

(d) Torque versus rotor angle characteristics.

In the case of the hydro turbine generator, the fault is removed in 0.466 s and 0.362 s in the case of the steam turbine generator. If the fault had been allowed to remain on the system slightly longer, the machines would have become unstable. i.e., they would either not have returned to synchronous speed after removal of the fault or slipped poles before returning to the synchronous speed.

When the fault is cleared, the system voltages are reapplied to the machine. Off-sets again occur in the phase currents, giving rise to the decaying 60-Hz oscillations in the rotor reference-frame variables and the instantaneous torque. The dynamic torque-angle characteristics shown in figure (b) and (d) yield a very lucid illustration of the fault and switching sequence and the return of the machine to its original operating condition after the fault is cleared.

The expression for the steady-state torque-angle characteristic for the hydro turbine generator is given by:

$$T_e = (52.1 \sin\delta + 12.5 \sin 2\delta) \times 10^6 \text{ N.m} \qquad ...(1)$$

For the steam turbine generator:

$$T_e = 3.05 \times 106 \sin\delta \text{ N.m} \qquad ...(2)$$

If these steady-state torque-angle characteristics are plotted on figures (b) and (d), respectively, they would pass through only the initial (or final) steady-state operating point. As in the case of a sudden change in its input torque, the instantaneous or average value of the dynamic or transient torque-angle characteristic differs markedly from the steady-state torque-angle characteristics.

In the case of a 3-phase fault close to a fully loaded machine, the circuit breakers would probably remove the machine from the system and reclosing would be prohibited because the machine would accelerate beyond the speed limits before it would be physically possible to reclose the circuit breakers.

1.4.3 Comparison of Actual and Approximate Transient Torque Characteristics

1.Comparison of actual and approximate transient torque-angle characteristics during a sudden change in input torque: First swing transient stability limit

In the studies involving a step increase in input torque, the machines were initially operating with essentially zero stator current and zero input torque. Hence, $\tilde{E}_a = \tilde{V}_{as}$ and $E'_q = \sqrt{2}V_S$. The steady-state torque-angle curve for the hydro unit and for the steam unit. For the hydro turbine generator, $X'_d = 0.3448$ and the transient torque-angle characteristic is:

$$T_e = (98.5\sin\delta - 20.5\sin 2\delta)\times 10^6\,N.m \quad ...(1)$$

For the steam turbine generator where $X'_d = 0.2591$ ohms, we obtain:

$$T_e = (6.92\sin 4 - 2.84\sin 24)\times 10^6\,N.m \quad ...(2)$$

Figures (a) and (b) show the approximate transient and steady-state torque-angle curves plotted along with the actual dynamic torque-angle characteristics obtained from the studies involving a step increase in input torque.

During the initial swing of the rotor, the dynamic torque-angle characteristic follows the approximate transient torque-angle curve more closely than the steady-state curve even though the approximation is rather crude, especially in the case of the steam turbine generator. As the transients subside, the actual torque-angle characteristic moves toward the steady-state torque-angle curve.

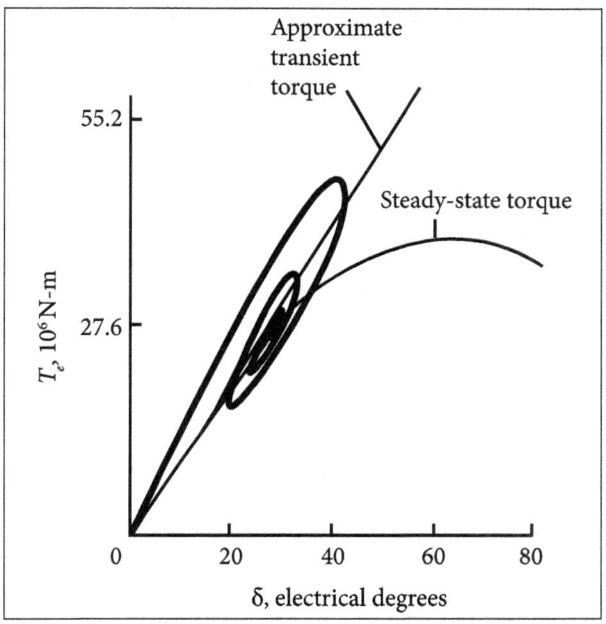

(a) Comparison of the dynamic torque-angle characteristic during a step increase in input torque from zero to rated with the calculated steady-state and approximate transient torque-angle characteristics: hydro turbine generator.

The approximate transient torque-angle characteristic is most often used along with the equal-area criterion to predict the maximum change in input torque possible without the machine becoming unstable (transient stability limit) rather than the dynamic performance during a relatively small step increase in input torque as portrayed in figures (a) and (b).

In order to compare the results obtained here with the transient stability limit, it is necessary to define the "first swing" transient stability limit as the maximum value of input torque which can be suddenly applied, and the rotor just returns to synchronous speed at the end of the first acceleration above synchronous speed. By trial and error, the transient stability limit for the hydro turbine generator was found to be 76.7×10^6 N.m and 5.2×10^6 N.m for the steam turbine generator.

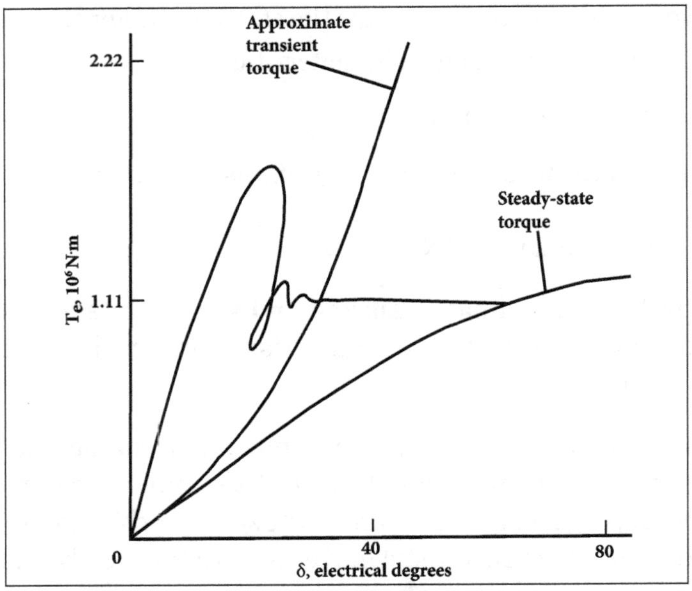

(b) Steam turbine generator with input torque stepped from zero to 50% rated.

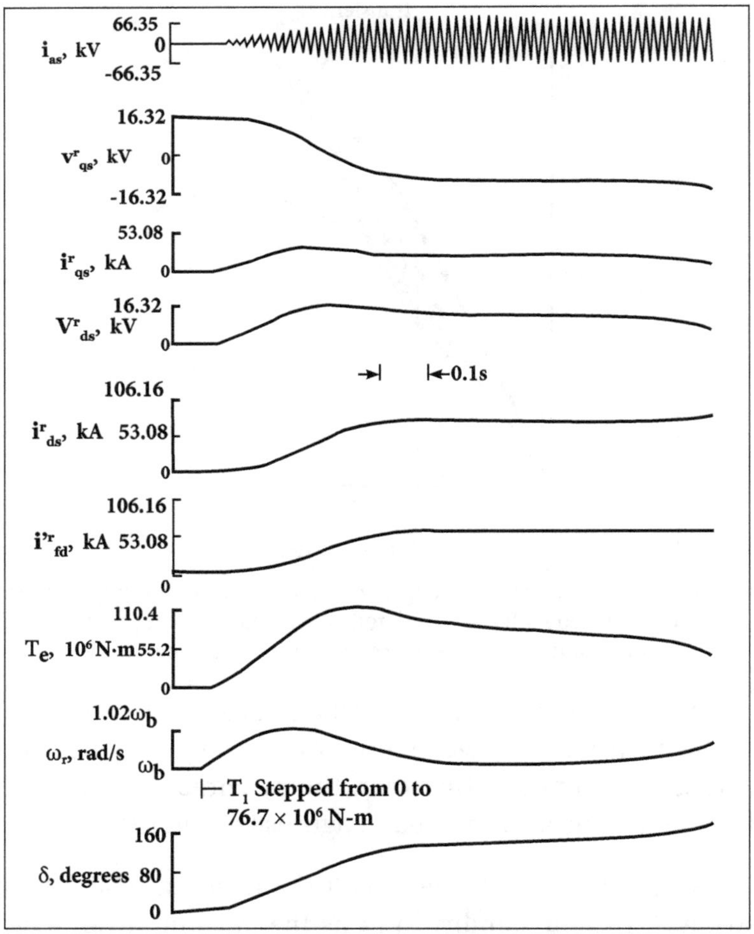

(c) Dynamic performance of a hydro turbine generator at the "first swing" transient stability limit.

The computer traces shown in figures (c) through (f) show the transient response of the machine variables with the step input torque equal to the value at the "first swing" transient stability limit. Figures (c) and (d) are for the hydro turbine generator and steam turbine generator, respectively, with the same variables. The machine flux linkages per second are shown in figure (e) for the hydro turbine generator and in figure (f) for the steam turbine generator.

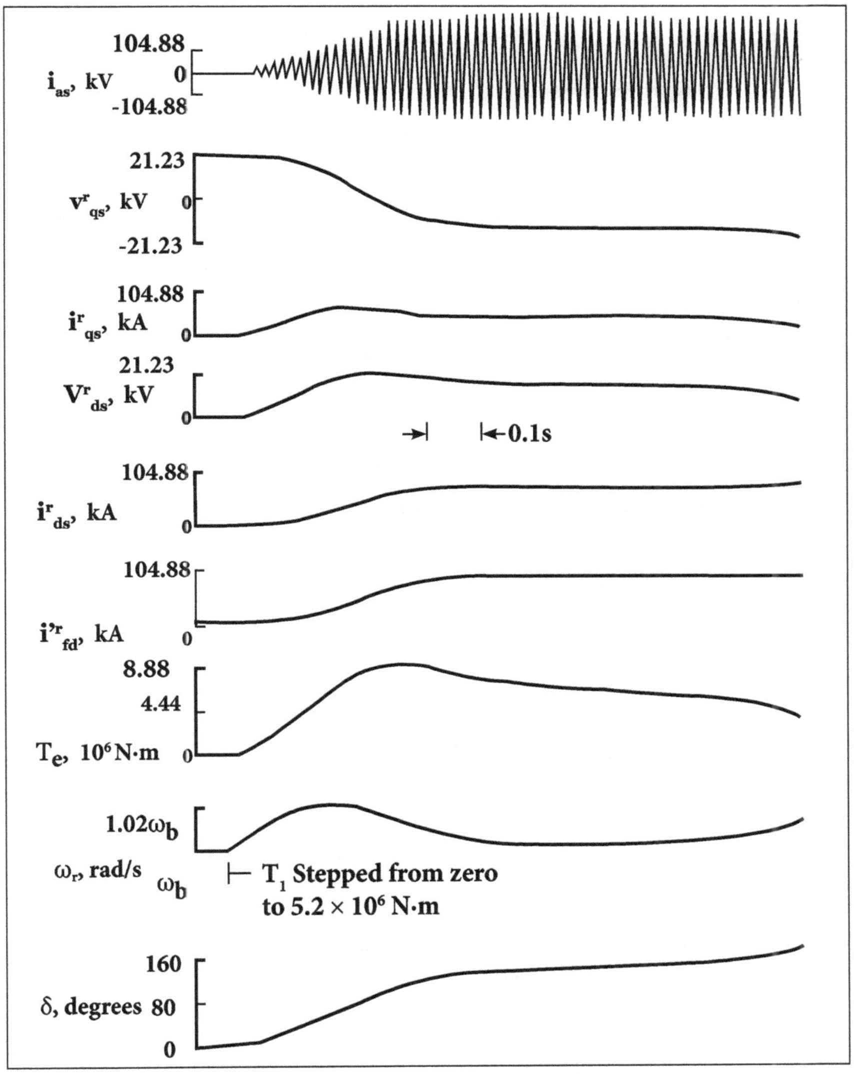

(d) Steam turbine generator.

In case of the hydro unit, the field flux linkages are relatively constant during the first swing of the machine varying approximately 17% from the original value (figure (e)). However, the field flux linkages for the steam unit (Figure (f)) vary approximately 40% during the first swing. This is due to the fact that the field circuit of the hydro unit has a higher reactance-to-resistance ratio than the field circuit of the steam unit. This observation casts doubt on the accuracy of the approximate transient torque-angle characteristics, especially for the steam unit, as we have already noted in figure (b). It is also important to note from figure (f) that the change in ψ^{Ir}_{kq1} is much less than that in the case of ψ^{Ir}_{fd}. The reactance-to-resistance ratios of the two circuits give an indication of this result. In particular, the reactance-to-resistance ratio of the field circuit of the steam unit is 1333 whereas this ratio for the kq1 winding is 1468.

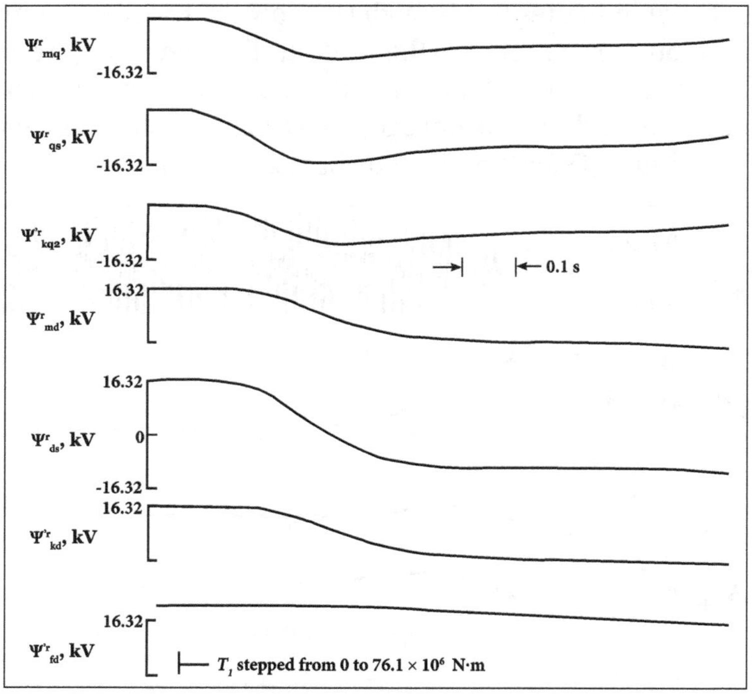

(e) Traces of flux linkages per second for a hydro turbine generator at the "first swing" transient stability limit.

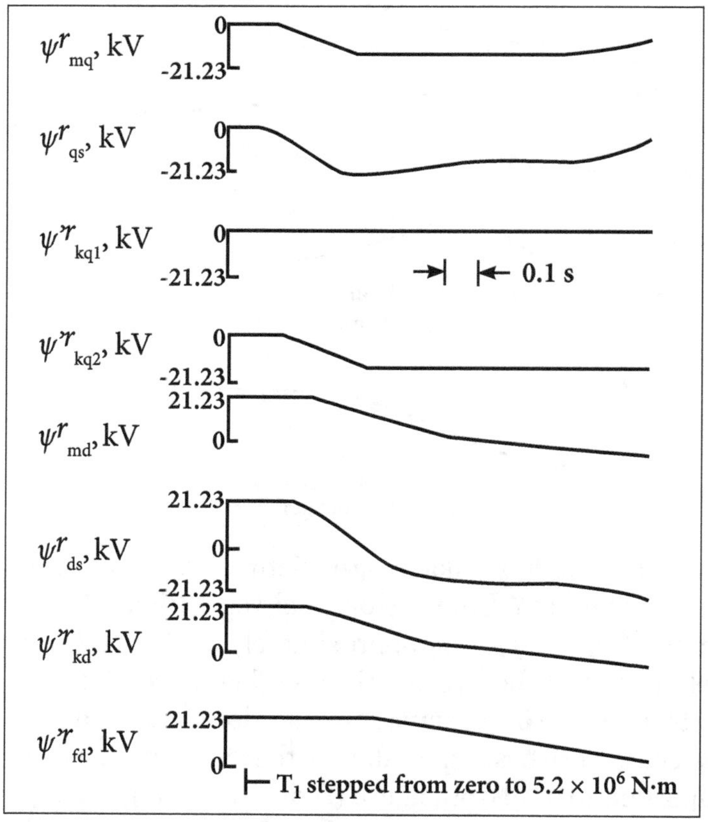

(f) Steam turbine generator.

The actual dynamic torque-angle characteristics obtained from the computer study are shown in Figures (g) and (h) with the approximate transient torque-angle curve calculated using constant flux linkages superimposed thereon. In the computer study the machines went unstable at the transient stability limit. Nevertheless, the transient stability limit has been defined here as the "first swing" transient stability limit in an attempt to duplicate the conditions for which the equal-area criterion is to predict.

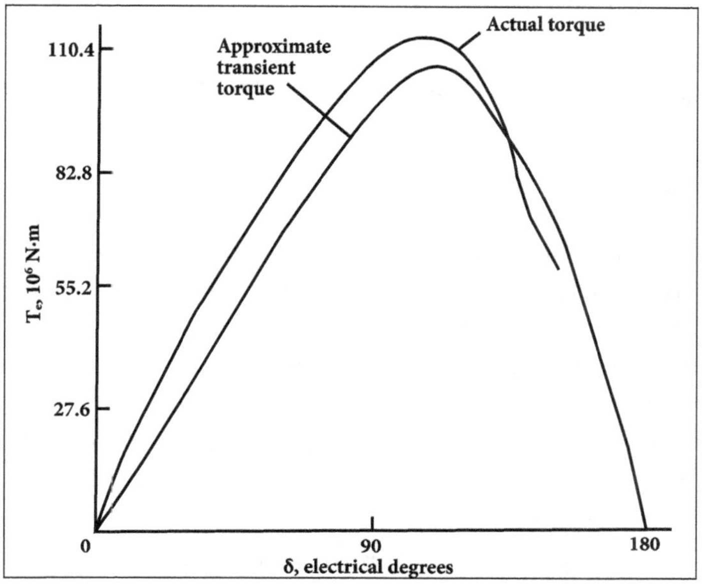

(g) Comparison of the dynamic torque-angle characteristic at the "first swing" transient stability limit with the approximate transient torque-angle curve: hydro turbine generator.

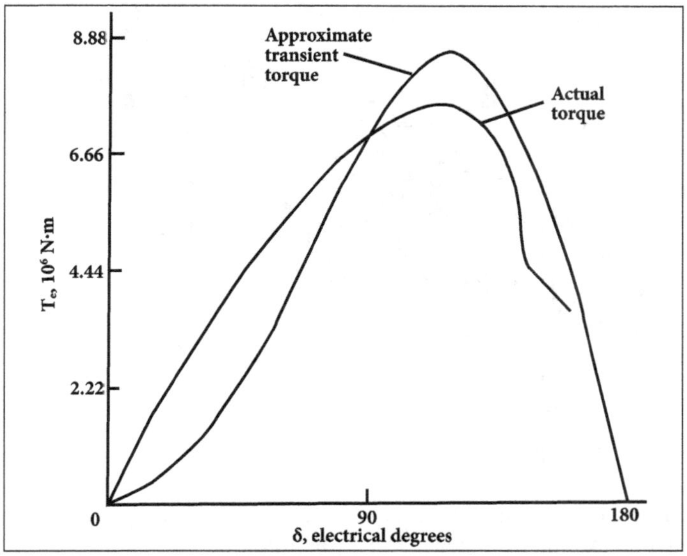

(h) Steam turbine generator.

The facility of the approximate transient torque-angle curve in predicting this transient stability must be deferred until we have described this criterion.

2. Comparison of Actual and Approximate Transient Torque-angle Characteristics During A 3-Phase Fault at the Terminals: Critical Clearing Time.

In the studies involving a 3-phase fault at the terminals, each machine is initially operating at rated conditions. In the case of the hydro turbine generator the steady-state torque-angle curve is given by (1), and for the steam turbine generator. For rated conditions, $E'_q = (1.16)\sqrt{\frac{2}{3}}20kV$ for the hydro unit and the approximate transient torque-angle characteristic is:

$$T_e = (114.3\sin\delta - 20.5\sin 2\delta)\times 10^6 \text{ N.m} \quad ...(1)$$

For the steam turbine generator, $E'_q = (1.09)\sqrt{\frac{2}{3}}26kV$ and the approximate transient torque-angle characteristic is:

$$T_e = (7.53\sin\delta - 2.84\sin 2\delta)\times 10^6 \text{ N.m} \quad ...(2)$$

Figures (a) and (b) show the approximate transient and steady-state torque-angle curves for the hydro and steam units, respectively, plotted along with the actual dynamic torque-angle characteristics shown previously in figures (b) and (d). It is important to note that the approximate transient and steady-state torque-angle curves both pass through the steady-state operating point. The flux linkages per second during a 3-phase fault and subsequent clearing are shown in Fig. 5.14-3 for the hydro turbine generator and in figure (d) for the steam turbine generator. The corresponding plots of voltages, currents, torque and speed and rotor angle are shown in figures (a) and (c):

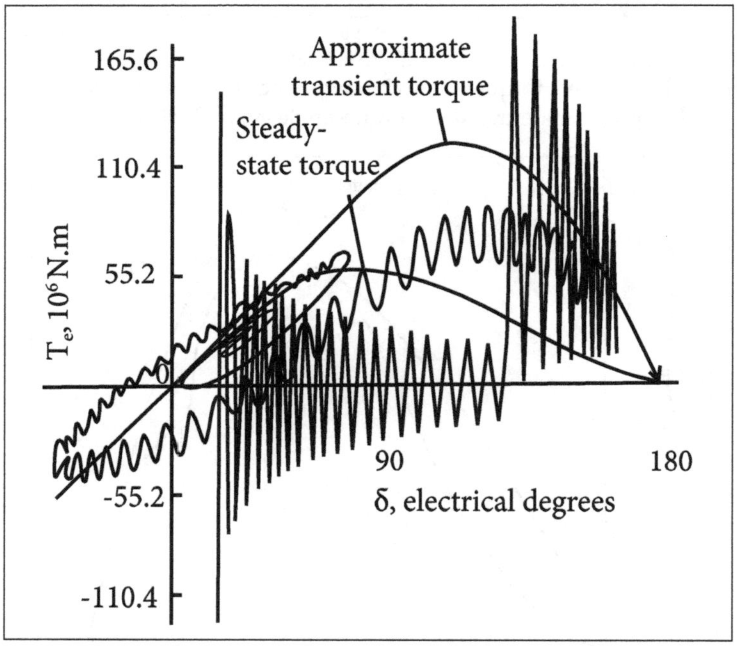

(a) Comparison of the dynamic torque-angle characteristic during a 3-phase fault with the calculated steady-state and approximate transient torque-angle characteristics: hydro turbine generator.

This limiting condition is commonly referred to as the critical clearing time, and the corresponding maximum rotor angle attained is called the critical clearing angle. In the case of the hydro turbine generator the critical clearing time and angle were found by trial and error to be 0.466 s and 123°. For the steam turbine generator these values were found to be 0.362 s and 128°. During the fault, the average value of the electromagnetic torque is essentially zero because the ohmic losses are small. The approximate transient torque-angle curve during this period is zero because the ohmic losses are neglected.

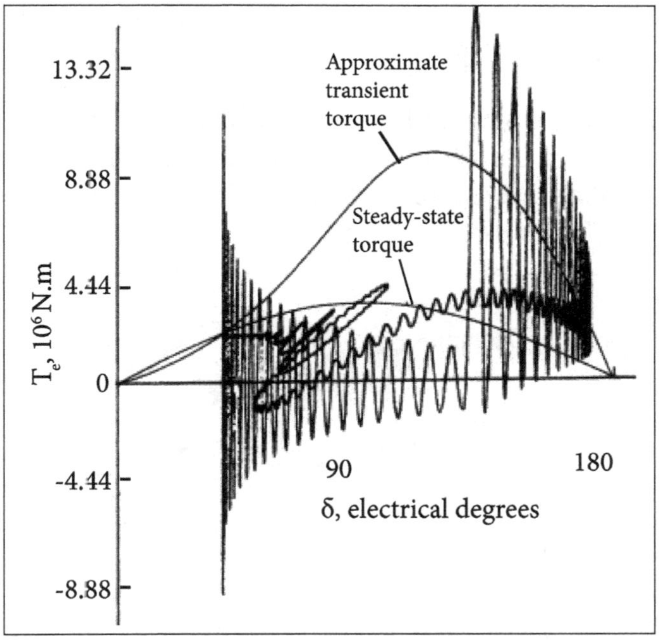

(b) Steam turbine generator.

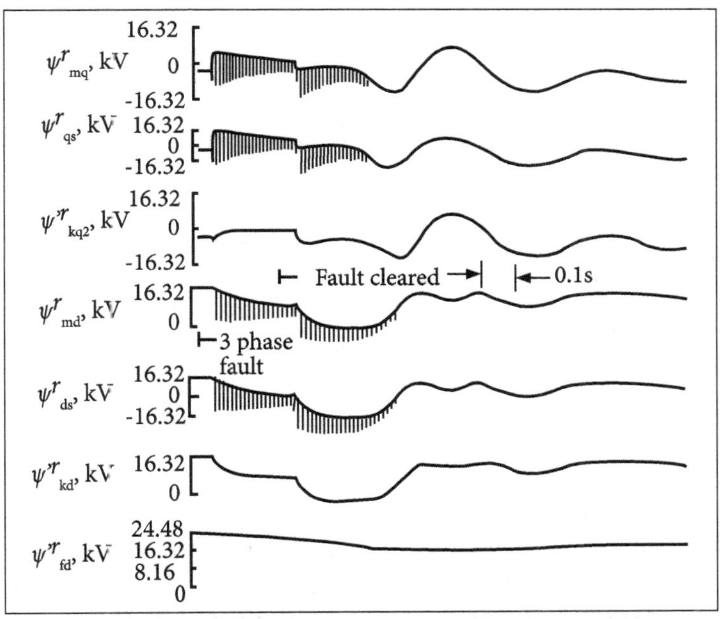

(c) Traces of flux linkages per second for a hydro turbine generator during a 3-phase fault at the terminals.

The approximate transient torque-angle curve plotted in figures (a) and (b) applies only before the 3-phase fault and after it is cleared. If the approximate transient torque-angle curve accurately portrayed the electromagnetic torque following the fault, the average of the instantaneous torque would traverse back and forth on this torque-angle characteristic until the initial operating condition was re-established.

The approximate characteristic appears to be adequate immediately following the fault when the rotor has reached the maximum angle. Thereafter, the approximation is quite inadequate especially in the case of the steam turbine generator (Figure (b)). This inaccuracy can of course be contributed primarily to the fact that the average value of the field flux linkages does not remain constant during and following the fault.

The change in field flux linkages is less in the case of the hydro unit, and the approximation is more accurate than that in the case of the steam unit. As in the case of a step increase in input torque, the damper winding flux linkages, 4/41, of the steam unit remains more nearly constant than any of the other machine flux linkages.

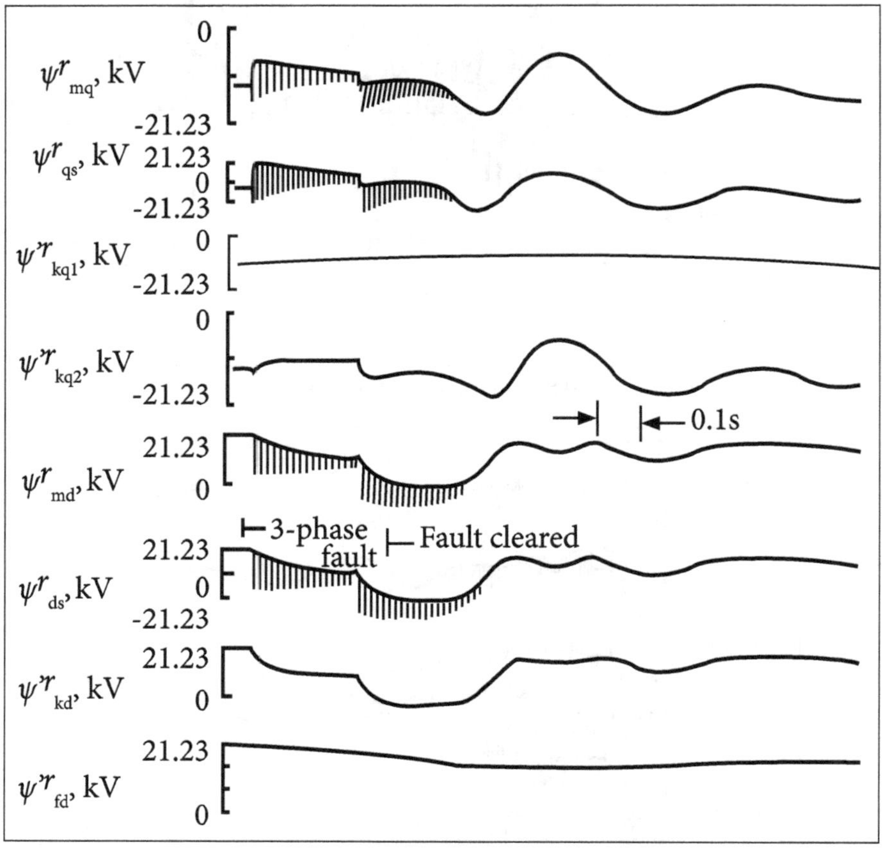

(d) Steam turbine generator.

Although the assumption of constant field flux linkages is by far the most common, there are refinements that can be made to yield the approximate transient torque-angle characteristic more accurate. in particular, it is sometimes assumed that the flux linkages of a q-axis damper winding also remains constant along with the field flux linkages. A voltage, different from E'_{q}, is then calculated behind a transient impedance.

2

Modeling of Synchronous Machines and Poly-Phase Induction Machines

2.1 Synchronous Machine: Introduction

Synchronous machines are AC machines which have a field circuit supplied by an external DC source. Synchronous machines consists of two major parts namely a stationary part and a rotating field system. In a synchronous generator, a DC current is applied to the rotor winding producing a rotor magnetic field.

The rotor is then driven by external means producing a rotating magnetic field, which induces a 3-phase voltage within the stator winding. Field windings are windings which produces the main magnetic field and armature windings are the windings where the main voltage is induced.

Types of Synchronous Machines

- Hydro generators: The generators that are driven by hydraulic turbines are called hydro generators. They run at lower speeds less than 1000 rpm.
- Turbo generators: These are the generators which are driven by steam turbines. These generators run at very high speed of 1500rpm or above.
- Engine driven Generators: These gets driven by IC engines. They run at a speed less than 1500 rpm. Hence the prime movers for the synchronous generators are Hydraulic turbines, Steam turbines or IC engines.

Construction of Single Phase Induction Motor

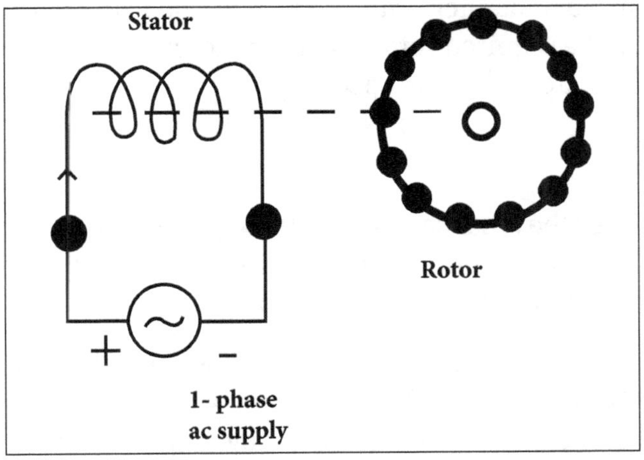

The main components of the Single Phase induction motor are stator and rotor. Stator is known to be the stationary part. Usually, the single phase alternating supply is given to the stator winding. The rotor is the rotating part of the motor. Rotor is connected to the mechanical load with the help of a shaft. A squirrel cage rotor is used here. It has a laminated iron core with many slots. Rotor slots are closed or semi-closed type.

The rotor windings are symmetrical and at the same time, it is short circuited. An air gap is there between the rotor and the stator. The most practical applications of this motor are in refrigerators, clocks, drills, pumps, washing machines etc. The stator winding in the 1Ø induction motor has two parts such as Main Winding and Auxiliary Winding. Normally, the Auxiliary winding is perpendicular to the main winding. In 1Ø induction motor, the winding with more turns is known as main winding while the other wire is called as auxiliary winding.

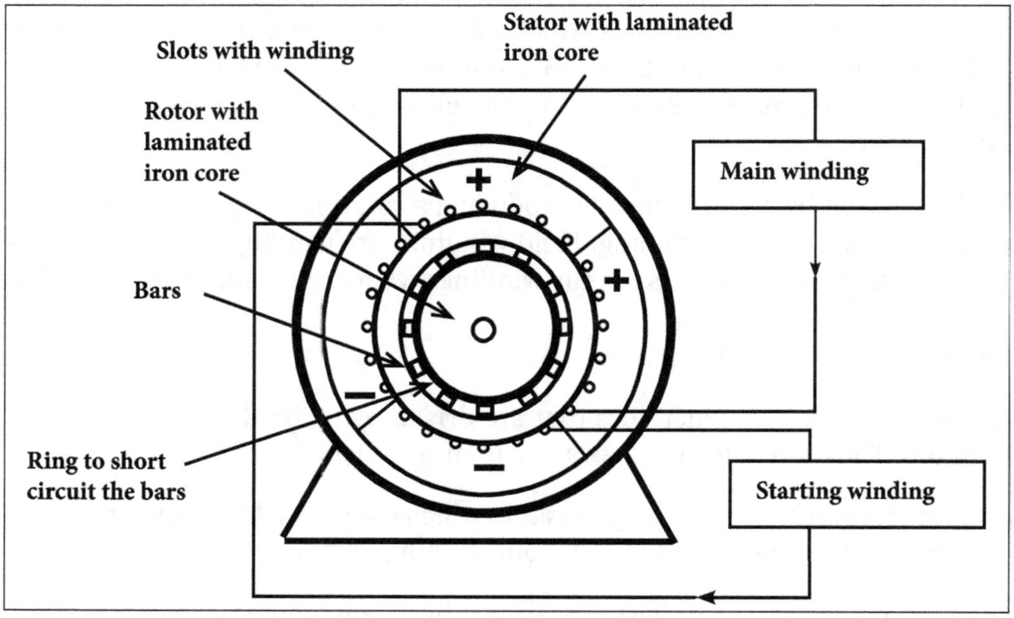

Construction of the single phase induction motor.

Principle of Operation

Single phase AC supply is given to the stator winding. Due to this, a magnetic field is produced which pulsates in the sinusoidal manner. After sometime, the field polarity reverses and the alternating flux cannot provide the required rotation of the motor. But if the motor is moved by external means, the motor will rotate with finite speed.

Circuit Model

The diagram below illustrates the cross section of a three phase, two pole cylindrical rotor synchronous machine. Coils aa', bb' and cc' represents the distributed stator windings producing sinusoidal mmf and flux density waves rotating in the air gap.

The reference directions for the currents are represented by using dots and crosses. The field winding ff' on the rotor also represents a distributed winding which produces sinusoidal mmf and flux density waves centered on its magnetic axis and rotating with the rotor.

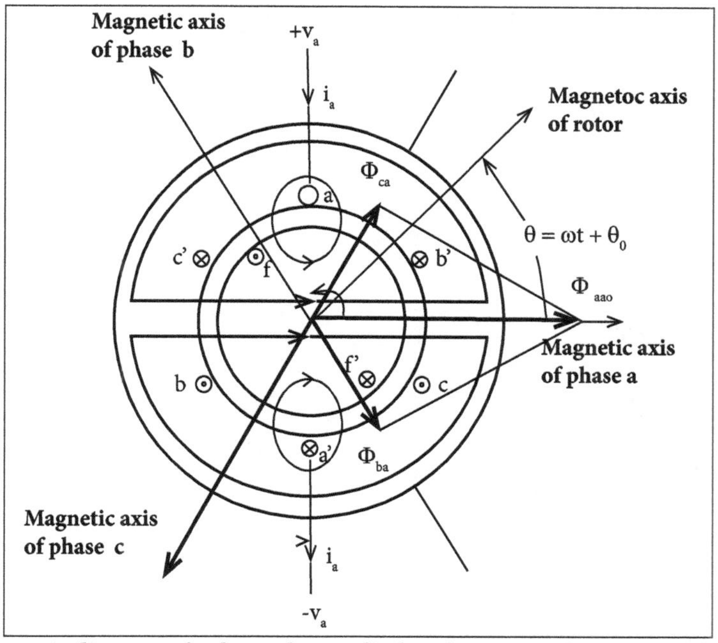

Schematic diagram of a three phase cylindrical rotor synchronous machine.

The electrical circuit equations for the three stator phase windings can be written by the Kirchhoff's voltage law as:

$$v_a = R_a i_a + \frac{d\lambda_a}{dt}$$

$$v_b = R_b i_b + \frac{d\lambda_b}{dt}$$

$$v_c = R_c i_c + \frac{d\lambda_c}{dt}$$

Where,

v_a, v_b and v_c are the voltages across the windings.

R_a, R_b and R_c are the winding resistances.

l_a, l_b and l_c are the total flux linkages of the windings of phases a, b and c, respectively.

For a symmetric three phase stator winding, we have:

$$R_a = R_b = R_c$$

The flux linkages of phase windings a, b and c can be expressed in terms of the self and mutual inductances as the following:

$$\ddot{e}_a = \ddot{e}_{aa} + \ddot{e}_{ab} + \ddot{e}_{ac} + \ddot{e}_{af} = L_{aa} i_a + L_{ab}\, i_b + L_{ac}\, i_c + L_{af} i_f$$

$$\ddot{e}_b = \ddot{e}_{ba} + \ddot{e}_{bb} + \ddot{e}_{bc} + \ddot{e}_{bf} = L_{ba} i_a + L_{bb}\, i_b + L_{bc}\, i_c + L_{bf} i_f$$

$$\ddot{e}_c = \ddot{e}_{ca} + \ddot{e}_{cb} + \ddot{e}_{cc} + \ddot{e}_{cf} = L_{ca} i_a + L_{cb}\, i_b + L_{cc}\, i_c + L_{cf} i_f$$

Where,

$$L_{aa} = L_{bb} = L_{cc} = L_{aao} + L_{a1}$$

$$L_{ab} = L_{ba} = L_{AC} = L_{ca} = -L_{aao}/2$$

$$L_{af} = L_{afm}\, \cos\theta$$

$$L_{bf} = L_{afm}\, \cos(\theta - 120°)$$

$$L_{cf} = L_{afm}\, \cos(\theta - 240°)$$

For a balanced three phase machine:

$$L_{aao} = \phi_{aao} / i_a$$

$$L_{al} = \phi_{al} / i_a$$

Where,

ϕ_{aao} is the flux that links all three phase windings.

ϕ_{al} is the flux that links only phase a winding.

$\theta = \omega t + \theta_o$.

When the stator windings are excited by balanced three phase currents, we have:

$$i_a + i_b + i_c = 0$$

The total flux linkage of phase a winding can be further written as:

$$\lambda_a = (L_{aao} + L_{al}) i_a - L_{aao}\, i_b / 2 - L_{aao}\, i_c / 2 + L_{afm}\, i_f \cos(\omega t + \theta_o)$$

$$= (L_{aao} + L_{al}) i_a - L_{aao}(i_b + i_c)/2 + L_{afm}\, i_f \cos(\omega t + \theta_o)$$

$$= (L_{aao} + L_{al}) i_a - L_{aao}\, i_a / 2 + L_{afm} i_f \cos(\omega t + \theta_o)$$

$$= (3L_{aao}/2 + L_{al}) i_a + L_{afm}\, i_f \cos(\omega t + \theta_o)$$

$$= L_s i_a + L_{afm}\, i_f \cos(\omega t + \theta_o)$$

Similarly, we can write:

$$\lambda_b = L_s\, i_b + L_{afm}\, i_f \cos(\omega t + \theta_o - 120°)$$

And $\lambda_c = L_s\, i_c + L_{afm}\, i_f \cos(\omega t + \theta_o - 240°)$

Where $L_s = 3L_{aao}/2 + L_{al}$ is known as the synchronous inductance.

In this way, the three phase windings are mathematically de-coupled and hence for a balanced three phase synchronous machine, we need to solve the circuit equation of one phase. Substituting the above expression of flux linkage into the circuit equation of phase a, we obtains:

$$v_a = R_a i_a + L_s \frac{di_a}{dt} + \frac{d\lambda_{af}}{dt}$$

In the steady state, the above equation can be expressed in terms of voltage and current phases as:

$$V_a = E_a + (R_a + j\omega L_s) I_a = E_a + (R_a + jX_s) I_a :$$

Where,

$X_s = \omega L_s$ is known as the synchronous reactance and

$E_a j \frac{\omega L_{afm} I_f}{\sqrt{2}} = j \frac{2\pi}{\sqrt{2}} f k_w N_{ph} \Phi_f = j4.44 f k_w N_{ph} \Phi_f$ is the induced emf phasor, nothing that

$$L_{afm} I_f = \lambda_{afm} = k_w N_{ph} \phi_f ,$$

Where,

I_f is the DC current in the rotor winding and φ_f the rotor magnetic flux in the air gap.

The above circuit equation was derived under the assumption that the phase current flows into the positive terminal, i.e., the reference direction of the phase current was chosen assuming the machine is a motor.

In case of a generator, where the phase current is assumed to flow out of the positive terminal, the circuit equation becomes:

$$V_a = E_a - (R_a + jX_s) I_a$$

The following circuit diagrams illustrate the per phase equivalent circuits of a round rotor synchronous machine in the motor and generator mode respectively.

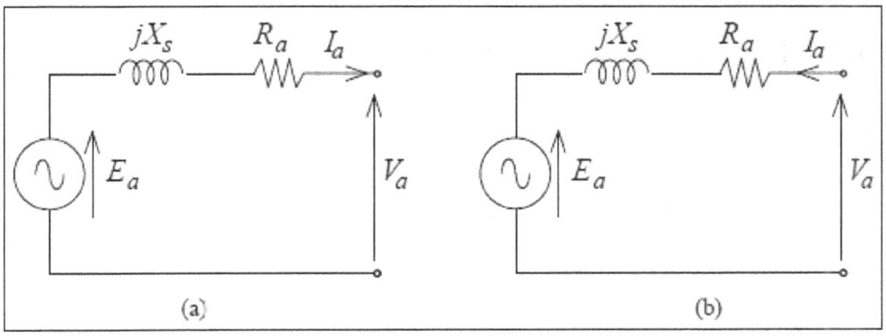

Synchronous machine per phase equivalent circuits in (a) generator and (b) motor reference directions.

2.1.1 Synchronous Machine Inductances Voltage Equations in the Rotor's dqo Reference Frame

Transformation to Rotor Reference Frame

Transform the stator quantities to rotor qdo reference frame, then the voltage equation has time-invariant coefficient. Rotor voltage equation already attached to q-axes and d- axes, qdo transformation only need to be applied to the stator winding.

For transformation, we denote:

$$v_{qdo} = T_{qdo}(\theta_r)vs(abc)$$

$$i_{qdo} = T_{qdo}(\theta_r)i_{s(abc)}$$

$$\wedge_{qdo} = T_{qdo}(\theta_r)\wedge_{s(abc)}$$

Where $T_{qdo}(\theta_r)$ is written as:

$$\left[T_{qdo}(\theta_r)\right]=\frac{2}{3}\begin{bmatrix} \cos\theta_r & \cos\left(\theta_r-\frac{2\pi}{3}\right) & \cos\left(\theta_r+\frac{2\pi}{3}\right) \\ \sin\theta_r & \sin\left(\theta_r-\frac{2\pi}{3}\right) & \sin\left(\theta_r+\frac{2\pi}{3}\right) \\ \frac{1}{2} & \frac{1}{2} & \frac{1}{2} \end{bmatrix}$$

Apply transformation $T_{qdo}(\theta_r)$ into stator voltage equation, the stator voltage equation becomes:

$$v_{qdo} = T_{qdo}\left(r_s i_{s(abc)}\right)+T_{qdo}\left(\frac{d}{dt}\wedge_{s(abc)}\right)=T_{qdo}r_s\left(T_{qdo}^{-1}i_{qdo}\right)+T_{qdo}\frac{d}{dt}\left(T_{qdo}^{-1}\wedge_{qdo}\right)$$

Transformation to Rotor Reference Frame

The first term of the transforming equation on the flux linkage:

$$T_{qdo}\left[\frac{d}{dt}T_{qdo}^{-1}\right]\Lambda_{qdo}=\omega_r\begin{bmatrix} 0 & 1 & 0 \\ -1 & 0 & 0 \\ 0 & 0 & 0 \end{bmatrix}\Lambda_{qdo}$$

The second term of the transforming equation on the flux linkage:

$$T_{qdo}T_{qdo}^{-1}\frac{d}{dt}\Lambda_{qdo}=\frac{d}{dt}\Lambda_{qdo}$$

The stator voltage equation of qdo rotor reference frame:

$$v_{qdo}=r_s i_{qdo}+\omega_r\begin{bmatrix} 0 & 1 & 0 \\ -1 & 0 & 0 \\ 0 & 0 & 0 \end{bmatrix}\Lambda_{qdo}+\frac{d}{dt}\Lambda_{qdo}$$

Flux Linkage in Terms of Winding Currents

Flux linkage Λ_{qdo} and qdo current can be obtained from stator quantities:

$$\Lambda_{s,qdo} = \left(T_{qdo} L_{ss} T_{qdo}^{-1}\right) i_{qdos} + \left(T_{qdo} L_{sr}\right) i_{qdor}$$

The stator qdo flux linkage equations:

$$\lambda_{qs} \left\{ L_{ls} + \frac{3}{2}\left(L_o - L_{ms}\right) \right\} i_q + L_{sfq} i_{sfq} + L_{skq} i_{skq}$$

$$\lambda_{ds} \left\{ L_{ls} + \frac{3}{2}\left(L_o - L_{ms}\right) \right\} i_d + L_{sfd} i_{sfd} + L_{skd} i_{skd}$$

$$\lambda_{os} + L_{ls} i_o$$

The rotor qdo flux linkage equations:

$$\lambda_{fd} = \frac{3}{2} L_{sfd} i_d + L_{ff} i_f + L_{fkd} i_{kd}$$

$$\lambda_{kd} = \frac{3}{2} L_{skd} i_d + L_{fkd} i_f + L_{kdkd} i_{kd}$$

$$\lambda_{fq} = \frac{3}{2} L_{sfq} i_q + L_{fqfq} i_{fq} + L_{fqkq} i_{kq}$$

$$\lambda_{kq} = \frac{3}{2} L_{skq} i_q + L_{fqkq} i_{fq} + L_{kqkq} i_{kq}$$

Voltage Equations in Rotor Reference Frame

Voltage equations of synchronous machine:

$$v_q = r_s i_q + \frac{d\lambda_q}{dt} + \lambda_d \frac{d\theta_r}{dt}$$

$$v_d = r_s i_d + \frac{d\lambda_d}{dt} - \lambda_q \frac{d\theta_r}{dt}$$

$$v_o = r_s i_o + \frac{d\lambda_o}{dt}$$

$$\dot{v}_{fd} = \dot{r}_{fd} i_{fd} + \frac{d\dot{\lambda}_{fd}}{dt}$$

$$\dot{v}_{kd} = \dot{r}_{kd} i_{kd} + \frac{d\dot{\lambda}_{kd}}{dt}$$

$$\dot{v}_{fq} = \dot{r}_{fq}\dot{i}_{fq} + \frac{d\dot{\lambda}_{fq}}{dt}$$

$$\dot{v}_{kq} = \dot{r}_{kq}\dot{i}_{kq} + \frac{d\dot{\lambda}_{kd}}{dt}$$

Flux linkage equations in terms of L and i is given by:

$$\lambda_q = L_q i_q + L_{mq} i'_{fq} + L_{mq} i'_{kq}$$

$$\lambda_d = L_d i_d + L_{md} i'_{fd} + L_{md} i'_{kd}$$

$$\lambda_o = L_{ls} i_o$$

$$\lambda'_{fd} = L_{md} i_d + L_{md} i'_{kd} + L_{fdfd} i'_{fd}$$

$$\lambda'_{kd} = L_{md} i_d + L_{md} i'_{fd} + L_{kdkd} i'_{kd}$$

$$\lambda'_{fq} = L_{mq} i_q + L_{fqfq} i'_{fq} + \dot{L}_{mq} i'_{kq}$$

$$\lambda'_{kq} = L_{mq} i_q + L_{mq} i'_{fq} + L_{kqkq} i'_{kq}$$

Where,

$$L_q = L_{ls} + L_{mq}$$

$$L_d = L_{ls} + L_{md}$$

Equivalent qdo Circuit in Rotor Reference Frame

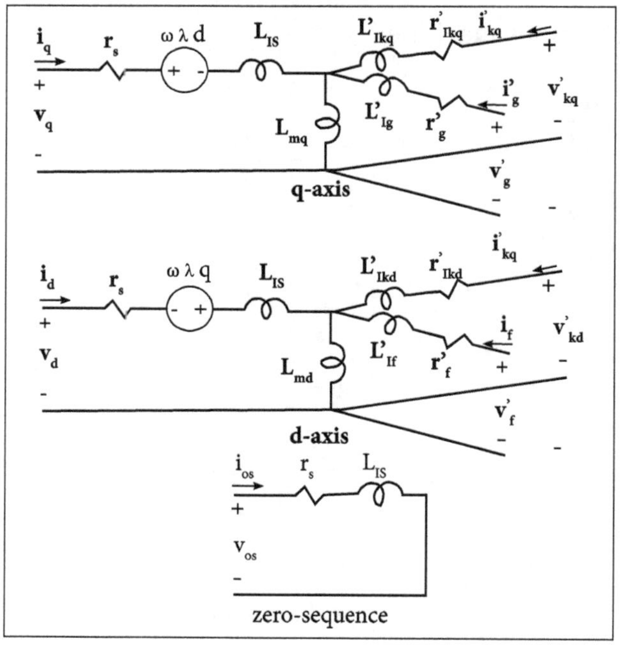

2.1.2 Electromagnetic Torque-current in Terms of Flux Linkages

Current Equations in Rotor Reference Frame

Current equations in terms of λ_s, q-axis is given by:

$$i_q = \frac{1}{L_{ls}}\left(\lambda_q - \lambda_{mq}\right)$$

$${}'_{fq} = \frac{}{{}'_{lfq}}\left(\lambda'_{fq} - \lambda_{mq}\right)$$

$$i'_{kq} = \frac{1}{L'_{lkq}}\left(\lambda'_{kq} - \lambda_{mq}\right)$$

Where $\lambda_{mq} = L_{mq}\left(i_q + i'_{fq} + i'_{kq}\right)$

$$\lambda_{mq} = \frac{L_{MQ}}{L_{ls}}\lambda_q + \frac{L_{MQ}}{L_{lfq}}\lambda'_{fq} + \frac{L_{MQ}}{L'_{lkq}}\lambda'_{kq}$$

$$\frac{1}{L_{MQ}} = \frac{1}{L_{ls}} + \frac{1}{L'_{lfq}} + \frac{1}{L'_{lkq}} + \frac{1}{L_{mq}}$$

Current equations in terms of λ_s, d-axis is written as:

$$i_d = \frac{1}{L_{ls}}\left(\lambda_d - \lambda_{md}\right)$$

$$i'_{fd} = \frac{1}{L'_{lfd}}\left(\lambda'_{fd} - \lambda_{md}\right)$$

$$i'_{kq} = \frac{1}{L'_{lkd}}\left(\lambda'_{kd} - \lambda_{md}\right)$$

Where $\lambda_{md} = L_{md}\left(i_d + i'_{fd} + i'_{kd}\right)$

$$\lambda_{md} = \frac{L_{MD}}{L_{ls}}\lambda_d + \frac{L_{MD}}{L'_{lfd}}\lambda'_f + \frac{L_{MQ}}{L'_{lkd}}\lambda'_{kd}$$

$$\frac{1}{L_{MD}} = \frac{1}{L_{ls}} + \frac{1}{L'_{lfd}} + \frac{1}{L'_{lkd}} + \frac{1}{L_{md}}$$

Matrix current equations in terms of flux linkage are given by:

$$\begin{bmatrix} i_d \\ i'_{fd} \\ i'_{kd} \end{bmatrix} = \begin{bmatrix} \left(1-\frac{L_{MD}}{L_{ls}}\right)\frac{1}{L_{ls}} & -\frac{L_{MD}}{L_{ls}L'_{lfd}} & -\frac{L_{MD}}{L_{ls}L'_{lkd}} \\ -\frac{L_{MD}}{L_{ls}L'_{lfd}} & \left(1-\frac{L_{MD}}{L'_{lfd}}\right)\frac{1}{L'_{ls}} & -\frac{L_{MD}}{L'_{lfd}L'_{lkd}} \\ -\frac{L_{MD}}{L_{ls}L'_{lkd}} & -\frac{L_{MD}}{L'_{lfd}L'_{lkd}} & \left(1-\frac{L_{MD}}{L'_{lfd}}\right)\frac{1}{L'_{lfd}} \end{bmatrix} \begin{bmatrix} \lambda_d \\ \lambda'_{fd} \\ \lambda'_{kd} \end{bmatrix}$$

The above matrix can be substituted into voltage and torque equation to obtain the math model of synchronous machine. The synchronous machine model derived from the above use flux linkages of rotor windings as state variables.

Electromagnetic Torque

Electromagnetic torque is obtained from the component of input power that transferred across the airgap.

Stating from the input power into the machine:

$$P_{in} = v_a i_a + v_b i_b + v_c i_c + v_{fd} i_{fd} + v_{fq} i_{fq}$$

Stator quantities transformed into qdo reference frame:

$$P_{in} = (3/2)(v_q i_q + v_d i_d) + 3v_o i_o + v_{fd} i_{fd} + v_{fq} i_{fq}$$

Eliminating the ohmic losses and rate of change of magnetic energy, we get:

$$P_{em} = (3/2)\omega_r(\lambda_d i_q - \lambda_q i_d)W$$

For a P-pole machine, $\omega_r = (P/2)\omega_{rm}$

$$P_{em} = (3/2)(P/2)\omega_{rm}(\lambda_d i_q - \lambda_q i_d W)$$

Dividing the P_{em} by ω_{rm}, the torque developed by P-pole machine is expressed as:

$$T_{em} = (3/2)(P/2)(\lambda_d i_q - \lambda_q i_d)\ \text{N.m}$$

Double Field Revolving Theory

According to double field revolving theory, any alternating quantity can be resolved into two components. Each component has magnitude equal to the half of the maximum magnitude of the alternating quantity. Both these component rotates in opposite direction to each other.

For example, a flux φ can be resolved into two components as $\frac{\phi_m}{2}$ and $-\frac{\phi_m}{2}$.

Each of these components rotates in opposite direction. i.e., if one $\phi_m/2$ is rotating in clockwise direction, then the other $\phi_m / 2$ rotates in the anticlockwise direction. When a single phase AC supply is given to the stator winding of single phase induction motor, it produces its flux of magnitude ϕ_m.

Accordingly the double field revolving theory, this alternating flux ϕ_m is divided into two components of magnitude $\phi_m / 2$. Each of these components will rotate in opposite direction with the synchronous speed N_s.

Let these two components of flux be referred as forward component of flux ϕ_f and backward component of flux ϕ_b. The resultant of these two component of flux at any instant of time gives the value of instantaneous stator flux at that particular instant.

$$\text{i.e.}\ \phi_r = \frac{\phi_m}{2} + \frac{\phi_m}{2} \quad \text{or} \quad \phi_r = \phi_f + \phi_b$$

Now at starting, both the forward and backward components of flux are exactly opposite to each other. Also both of these components of flux are equal in magnitude. So they cancel each other and so the net torque experienced by the rotor at starting is zero. So the single phase induction motors are not self-starting motors.

When the stator winding carries a sinusoidal current, a sinusoidal space distributed mmf, whose peak or maximum value pulsates with time, is produced in the air gap. This sinusoidally varying flux (φ) is the sum of two rotating fluxes or fields; the magnitude of flux is equal to half the value of the alternating flux (φ/2).

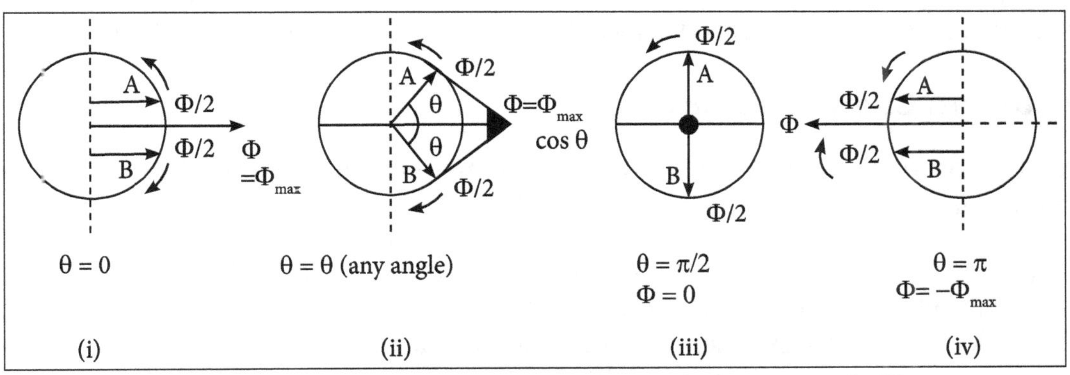

Position of the pulsating and rotating in fluxes with change in angle (θ).

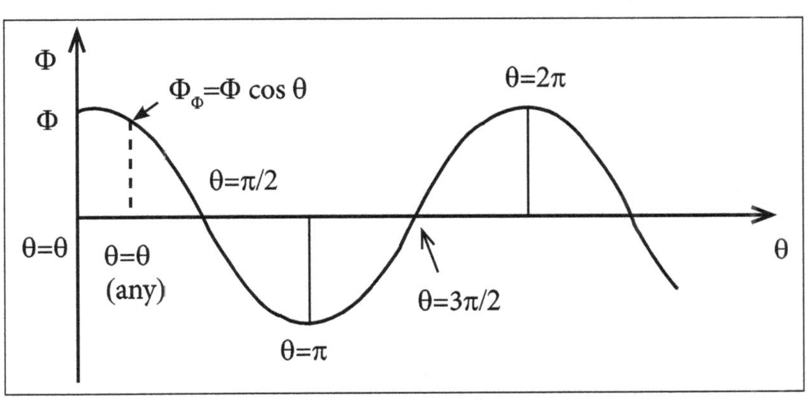

Pulsating (sinusoidal) flux as a function of space angle (θ).

Both the fluxes rotate synchronously at the speed, (n_s=2.f/p) in opposite directions. This is shown in the above figure. The first set of figures (i-iv) show the resultant sum of the two rotating fluxes or fields as the time axis (angle) is changing from $\theta = 0°$ to π (180°). The flux or field rotating at synchronous speed in the anticlockwise direction, i.e., the same direction as that of the motor (rotor) taken as positive induces emf (voltage) in the rotor conductors.

The rotor is a squirrel cage one with bars short-circuited via end rings. The current flows in the rotor conductors and electromagnetic torque is produced in the same direction is termed as positive (+ve). The other part of flux or field rotates at the same speed in the opposite (clockwise) direction taken as negative.

The torque produced by this field is negative (-ve) as it is in the clockwise direction same as that of the direction of rotation of this field. Two torques are in the opposite direction and the resultant torque is the difference of the two torques produced.

If the rotor is stationary, the slip due to forward (anticlockwise) rotating field is s_f = 0.1. The two torques are equal and opposite to the resultant torque is 0.0 (zero). So there is no starting torque in a single-phase IM. Therefore, S_b=0.1.

But if the motor (rotor) is started or rotated in the anticlockwise (forward) direction, the forward torque is more than the backward torque with the resultant torque now being positive.

The motor accelerates in the forward direction with the forward torque being more than the backward torque. The resultant torque is thus positive as the motor rotates in the forward direction. The motor speed is decided by the load torque supplied including the losses. Therefore, the rotor will stay at rest. So due to this effect, we have to find another method to start the motor.

Speed Current Characteristics

We have a proportional relationship between the counter emf and speed of the motor. If the value of the counter emf is large, then the value of speed is more.

From the phasor diagram, it is known that the counter emf is equal to the difference between the terminal voltage and the voltage drops. Hence, if the current causes higher voltage drops, then the generated back emf will be less. Therefore, the speed of the motor is less.

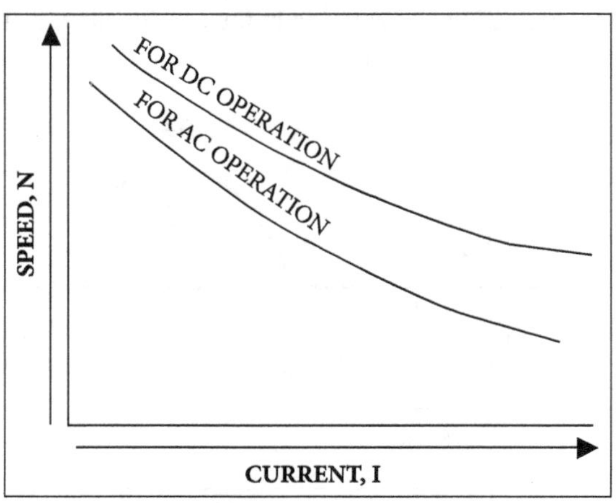

Speed - Current Characteristics.

In case of DC series motor, we have high value of counter emf because the value of voltage drop is small. The voltage drop is due to resistive drops mainly. Therefore, the value of voltage drop is low.

In case of AC series motor, we have a low value of counter emf because the value of voltage drop is large. The voltage drop is due to the resistive drops and reactance drop. Therefore, the value of voltage drop is high. It means the speed current characteristics curve for the DC series are less dropping than the AC series motor.

Torque Current Characteristics

After neglecting the small value of phase angle and saturation effect, torque is directly proportional to the value of square of the current. Therefore, the variation of torque with the current can plotted as shown in the below figure:

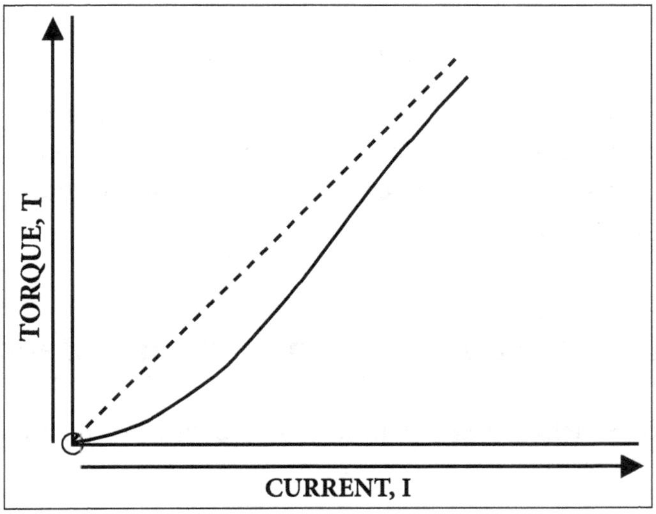

Torque - Current Characteristics.

Torque Speed Characteristics

The torque speed characteristics are plotted as shown in the below figure:

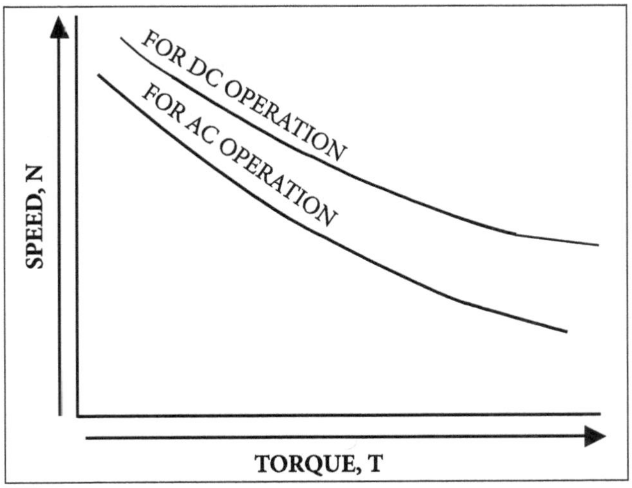

Torque Speed Characteristics.

Power Output Characteristics

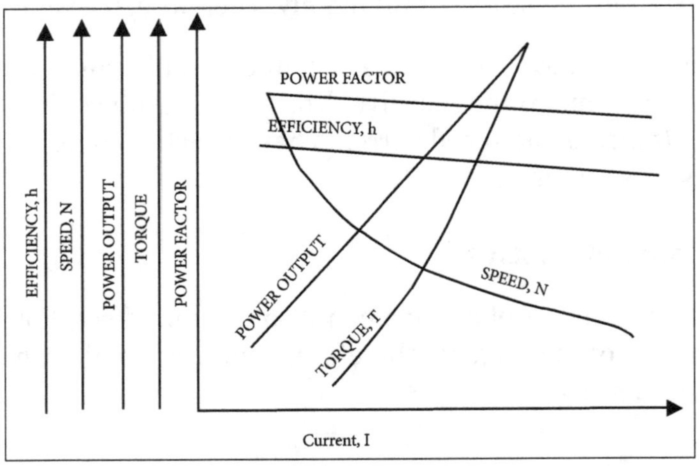

Power Output Characteristics.

The mechanical output power developed by the AC series motor can be calculated by the product of the counter emf and current. The value of mechanical power developed is directly proportional to the value of the current, if we neglect the decrement in value of the counter emf. The counter emf slightly decreases with increase in current.

2.2 Simulation of Three Phase Synchronous Machine

The winding equation of Synchronous machine model could use terminal voltage as input and currents as output.

Main inputs: Stator abc phase voltage, excitation voltage to field winding, applied mechanical torque (load).

Main outputs: Stator abc phase current.

Modeling Procedure

Stator winding voltage must be transformed to qdo reference frame attached to the rotor.

Step I:

$$v_q^s = \frac{2}{3}v_a - \frac{1}{3}v_b - \frac{1}{3}v_c$$

$$v_d^s = \frac{1}{\sqrt{3}}(v_c - v_b)$$

$$v_o = \frac{1}{3}(v_a + v_b + v_c)$$

Step II:

$$v_q^r = v_q^s \cos\theta_r(t) - v_d^s \sin\theta_r(t)$$

$$v_d^r = v_a^s \sin\theta_r(t) + v_d^s \cos\theta_r(t)$$

Where,

$$\theta_r(t) = \int_o^t \omega_r(t)dt + \theta_r(o) \qquad \text{elect.rad.}$$

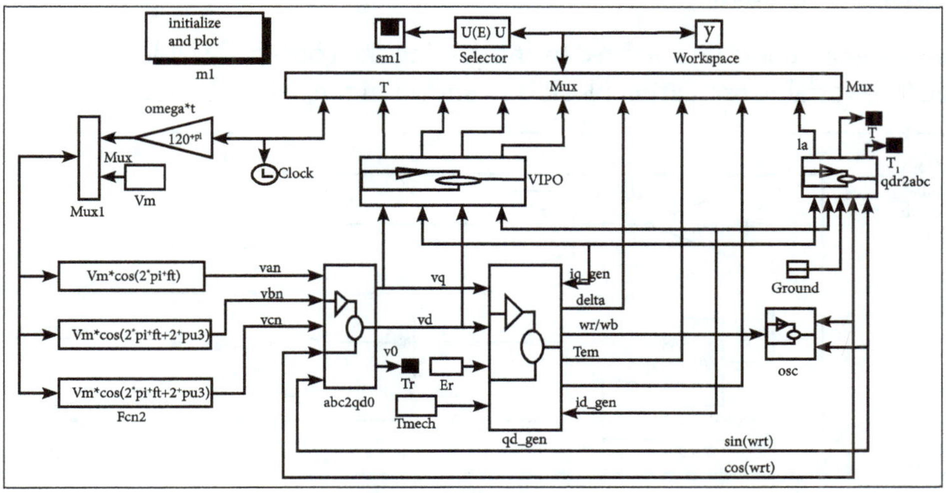

Overall Diagram.

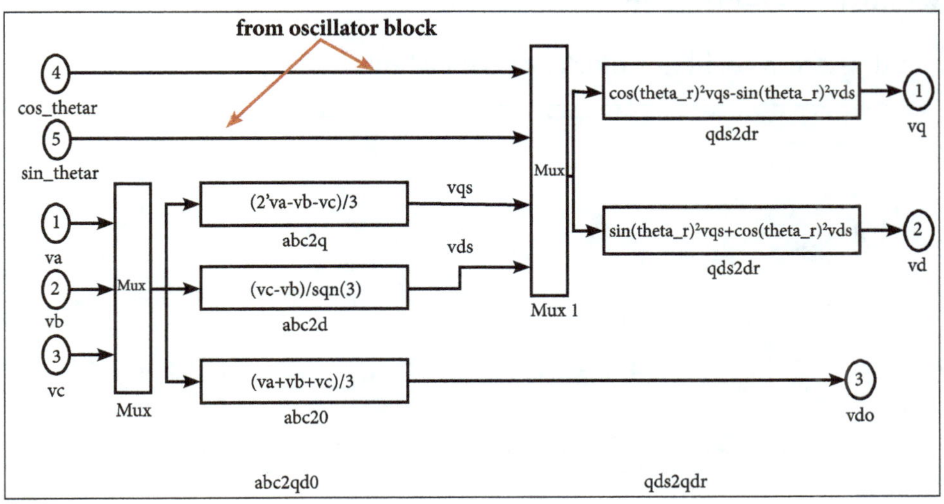

abc to rotor qdo block.

Variable-frequency Oscillator

The equation of the oscillator is given by:

$$\frac{d^2 y_1}{dt^2} = -\omega^2 y_1$$

Convert 2nd-order equation to two 1st-order equation as:

$$y_2 = \frac{1}{\omega}\frac{dy_1}{dt}, \quad \frac{dy_2}{dt} = -\omega y_1$$

Rewrite the above equations into integral forms as:

$$y_2 = -\omega\int y_1\,dt, \quad y_1 = \omega\int y_2\,dt,$$

Assume y_1 is sinωt, y_2 is cosωt.

The cosωt, sinωt signals could be obtained by integrating each other. Variable y_1, y_2 could be obtained by variable ωt. The variable oscillation block is as shown in the below figure:

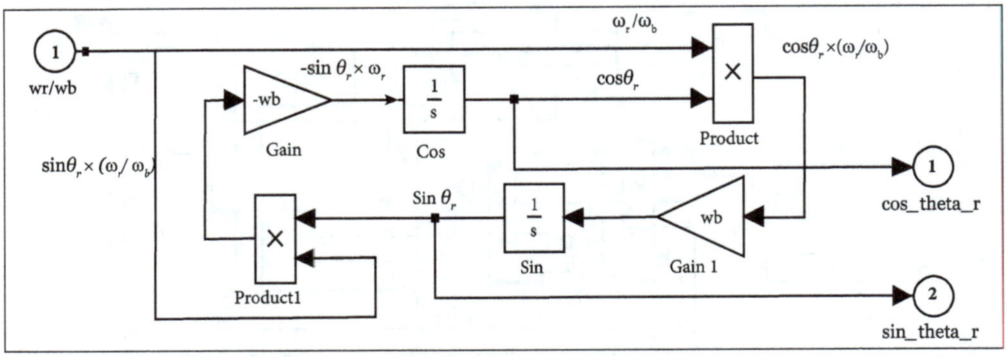

Simulink Result of Oscillator

Plot result: y_1 and y_2 (from oscillator block), ω is constant.

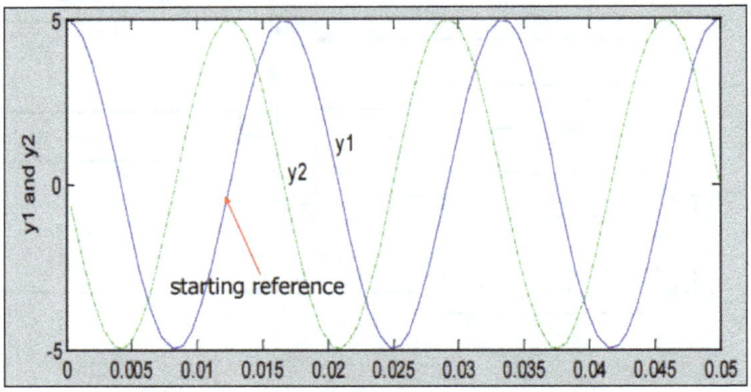

Terminal Voltage and Flux Linkage Equations

Modeling Procedure

- Stator: Relate terminal voltages (input variables) with the flux linkage of stator and rotor windings.
- Rotor: In this case, only flux linkages of field winding in d-axis and a pair of damper windings in d-axes and q-axes are related together.

$$\Psi_q = \omega_b \int \left\{ v_q - \frac{\omega_r}{\omega_b} \Psi_d + \frac{r_s}{x_{ls}} \left(\Psi_{mq} - \Psi_q \right) \right\} dt$$

$$\Psi_d = \omega_b \int \left\{ v_d - \frac{\omega_r}{\omega_b} \Psi_q + \frac{r_s}{x_{ls}} \left(\Psi_{md} - \Psi_d \right) \right\} dt$$

$$\Psi_o = \omega_b \int \left\{ v_o - \frac{r_s}{x_{ls}} \Psi_o \right\} dt$$

$$\Psi'_q = \frac{\omega_b r'_{kq}}{x'_{lkq}} \int \left(\Psi_{mq} - \Psi'_{kq} \right) dt$$

$$\Psi'_{kd} = \frac{\omega_b r'_{kd}}{x'_{lkd}} \int \left(\Psi_{md} - \Psi'_{kd} \right) dt$$

$$\Psi'_{fd} = \frac{\omega_b r'_{kq}}{x'_{md}} \int \left\{ E_f + \frac{x_{md}}{x'_{lf}} \left(\Psi_{md} - \Psi'_{fq} \right) \right\} dt$$

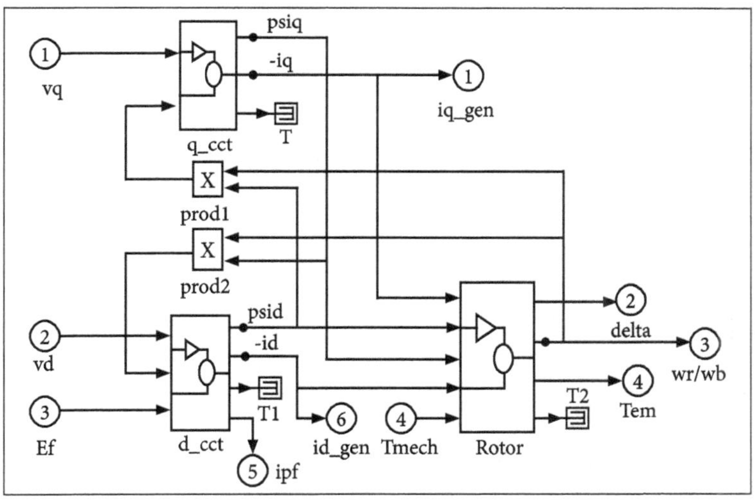

qdo stator and rotor block.

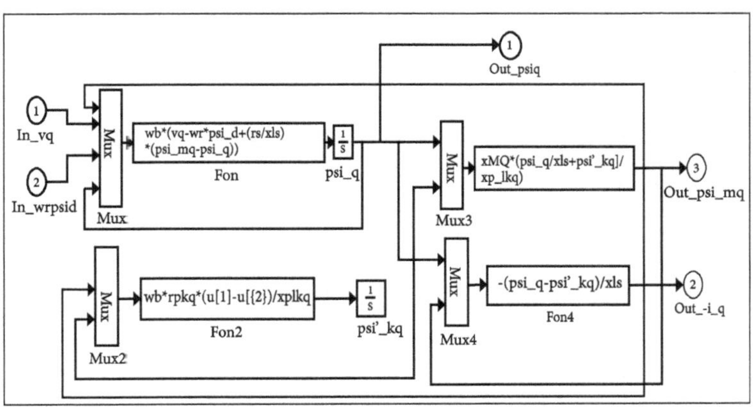

q-axis stator block.

In the figure:

Rotor flux linkage:

$$\Psi_{mq} = x_{mq}\left(i_q + i'_{kq}\right)$$

$$\Psi_{md} = x_{mq}\left(i_d + i'_{kq} + i'_{fd}\right)$$

$$E_f = x_{md}\frac{v'_f}{r'_f}$$

$$\Psi'_{fd} = x'_{lf}i'_{fd} + \Psi_{md}$$

$$\Psi'_{kd} = x'_{lkd}i'_{kd} + \Psi_{md}$$

$$\Psi'_{kq} = x'_{lkq}i'_{kq} + \Psi_{md}$$

Stator flux linkage:

$$\Psi_q = x_{ls}i_q + \Psi_{mq}$$

$$\Psi_d = x_{ls}i_d + \Psi_{md}$$

$$\Psi_o = x_{ls}i_o$$

Winding Currents

Modeling Procedure

Relate flux linkages with winding currents.

Stator winding current:

$$i_q = \frac{\Psi_q - \Psi_{mq}}{x_{ls}}$$

$$i_d = \frac{\Psi_d - \Psi_{md}}{x_{ls}}$$

Rotor winding current:

$$i'_{kd} = \frac{\Psi'_{kd} - \Psi_{md}}{x'_{lkd}}$$

$$i'_{kq} = \frac{\Psi'_{kq} - \Psi_{mq}}{x'_{lkq}}$$

$$i'_{fd} = \frac{\Psi'_{fd} - \Psi_{md}}{x'_{lfd}}$$

Modeling Procedure (dq0 to abc Transformation)

Transform qd0 rotor frame currents to stationary frame:

$$i_q^s = i_q^r \cos\theta_r(t) + i_d^r \sin\theta_r(t)$$

$$i_d^s = -i_q^r \sin\theta_r(t) + i_d^r \cos\theta_r(t)$$

Transform stationary qd0 currents to abc currents:

$$i_a = i_q^s + i_o$$

$$i_b = -\frac{1}{2} i_q^s - \frac{1}{\sqrt{3}} i_d^s + i_o$$

$$i_c = -\frac{1}{2} i_q^s + \frac{1}{\sqrt{3}} i_d^s + i_o$$

Torque Expression

Electromechanical torque developed with P-poles in motoring or generating:

$$T_{em} = \frac{P_{em}}{\omega_{rm}} = \frac{3}{2}\frac{P}{2}\left(\lambda_d i_q - \lambda_q i_d\right) \quad N \cdot m$$
$$= \frac{3}{2}\frac{P}{2\omega_b}\left(\psi_d i_q - \psi_q i_d\right) \quad N \cdot m$$

Where T_{em} is positive (+) for motoring and negative(-) for generating.

The net accelerating torque:

$$T_{em} + T_{mech} - T_{damp} = J\frac{d\omega_{rm}(t)}{dt} = \frac{2J}{P}\frac{d\omega_r(t)}{dt} = \frac{2J}{P}\frac{d\left(\omega_r(t) - w_e\right)}{dt}$$

$$\omega_r(t) - w_e = \frac{P}{2J}\int_o^t \left(T_{em} + T_{mech} - T_{damp}\right) dt$$

Rotor Block

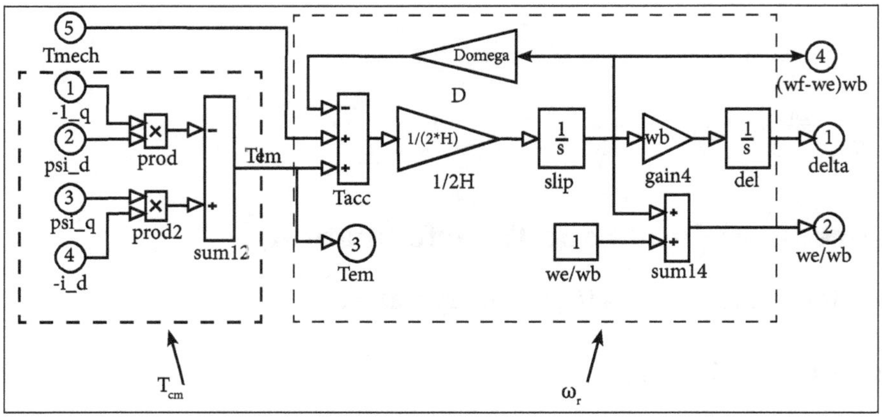

Per-Unit Expression for Torque and Speed

Per-unit Variables

$$V_b = \frac{\sqrt{2}}{\sqrt{3}} V_{line\text{-}to\text{-}line}, \quad I_b = \frac{2S_b}{3V_b}, \quad Z_b = \frac{V_b}{I_b}, \quad T_b = \frac{S_b}{\omega_{bm}}, \quad \omega_{bm} = \omega_b \frac{2}{P}$$

$$T_{em(pu)} + T_{mech(pu)} - T_{damp(pu)} = \frac{1}{T_b} \frac{2J}{P} \frac{d\omega_r(t)}{dt} pu$$

$$= 2H \frac{d\{(\omega_r(t) - w_e)/\omega_b\}}{dt} pu,$$

Where,

$$H = \frac{1}{2} J \omega_{bm}^2 S_b$$

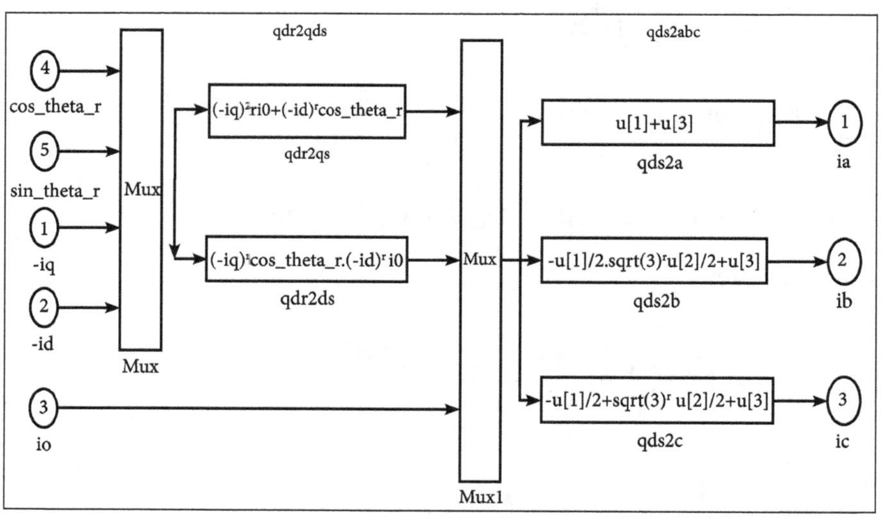

qdo rotor frame to abc block.

VIPQ block

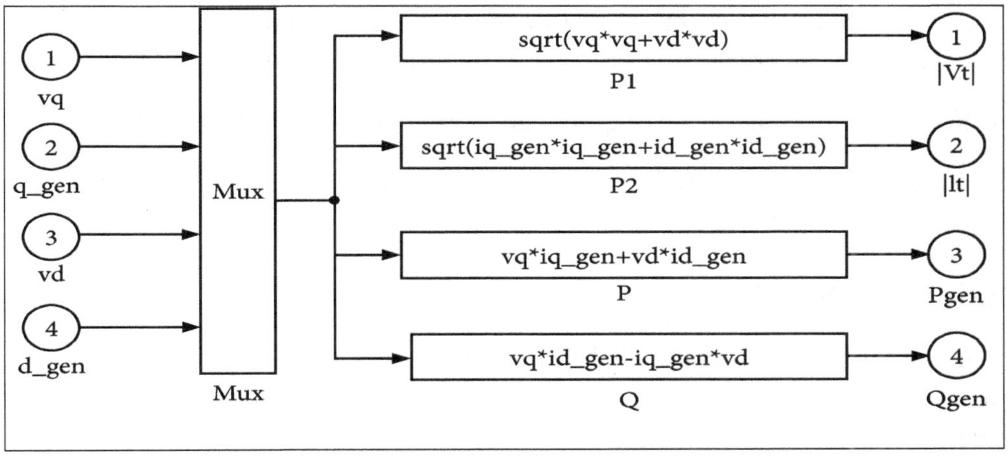

$$|V_t| = \sqrt{V_{qs}^2 + V_{ds}^2} \quad P = \text{Re}\left\{(v_q - jv_d)(i_q - ji_d)^*\right\} = v_q i_q + v_d i_d$$

$$|I_t| = \sqrt{I_{qs}^2 + I_{ds}^2} \quad Q = \text{Im}\left\{(v_q - jv_d)(i_q - ji_d)^*\right\} = v_q i_d - v_d i_q$$

2.3 Modeling of PM Synchronous Motor

A permanent magnet synchronous motor (PMSM) uses permanent magnets embedded in steel rotor to create a constant magnetic field. The stator carries windings connected to an AC supply to produce a rotating magnetic field. At the synchronous speed, the rotor poles lock to the rotating magnetic field. Permanent magnet synchronous motors are similar to the brushless DC motors. Because of the constant magnetic field in the rotor these cannot use the induction windings for starting. These motors require a variable-frequency power source to start.

The main difference between a permanent magnet synchronous motor and an asynchronous motor is the rotor. Some studies seem to indicate that the NdFeB permanent magnet synchronous motors are around 2% more efficient than the highest-efficiency (IE_3) asynchronous motors using same stator laminations and similar variable-frequency speed controllers.

Permanent magnet synchronous motors (PMSM's) are an attractive solution for servo drives in the kW-range. They are being used in a wide range of applications, from general purpose line-start pump or fan drives to high performance machine tool servos, robotics, aerospace generators and actuators and electric vehicles.

Materials for PMSM

The desirable properties of the materials used for the construction of permanent magnets in PMSM are:

- High flux density.

- High coercivity.

Certain "rare earth" magnetic materials satisfy these requirements. Samarium-Cobalt, with high flux density (up to 1T) and very large coercive force (upto 7000 A/cm), is one of the materials used.

These days, Neodymium-Iron-Boron magnets are being used for their low cost.

Further, the magnets made of these materials require less space and with proper design, there is no danger of accidental demagnetization through short circuit.

Classification of PMSM

In medium and small size motors, DC field can be produced by permanent magnets. The advantages are elimination of field copper loss, high power density, lower rotor inertia and more robust construction of rotor. The PMSM can be classified as:

1. Surface mounted type

There are two variations in the surface mounted PMSM.

a. Projecting type - Type in which the magnets project from the surface of the rotor as shown in the below figure:

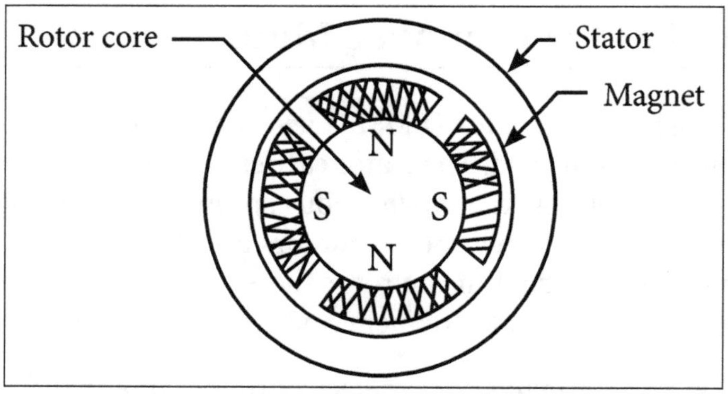

Projecting surface mounted PMSM.

b. Insert type - Type in which the magnets are inserted into the rotor providing a smooth rotor surface as shown in the below figure:

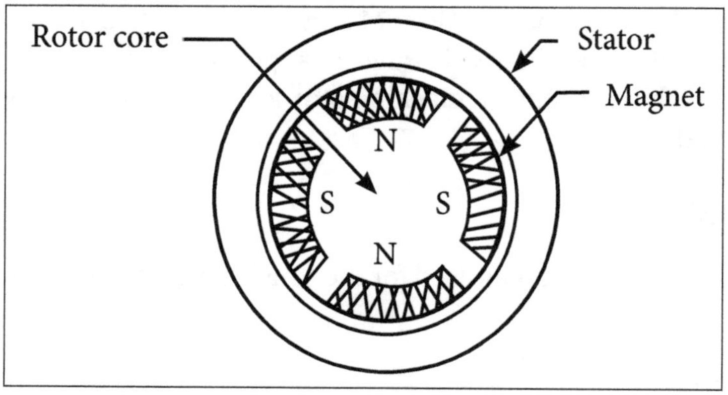

Insert magnet PMSM.

Projecting type: Mounting the magnets on the surface of the rotor is the simplest and cost effective method for constructing a PMSM. Epoxy glue is used to fix the magnets to the rotor surface in the projecting and insert surface mounted PMSM. Therefore, these constructions are less robust compared to interior magnet type rotor and are not suited for high speed applications.

As the relative permeability of the magnet is approximately equal to that of air, the projecting surface mounted motor has almost uniform airgap. This results in equal direct and quadrature axis inductances. The motor torque is therefore derived from the magnet flux only and there is no reluctance torque as might have been expected from the salient magnet construction.

The projecting magnets present a large airgap in the magnetic circuit, so that the inductance in the machine is small. The lower value of inductance gives rise to a smaller stator time constant. Also the surface mounted designs tend to yield smaller rotor diameter with lower inertia. This enables the machine to exhibit good dynamic performance. Such drives are suitable for high performance machine tool feed drives and industrial robots.

Insert Type: The insert PMSM have magnets that span less than a pole pitch, mounted into the rotor surface with an iron rotor tooth filling the space between the magnets. This design makes the machine mechanically more robust, the inductance is much larger compared with the projecting surface mounted PMSM.

The quadrature inductance is larger than the direct axis inductance, owing to the magnet being equivalent to an airgap in the direct axis. This saliency gives rise to a significant reluctance torque which adds to the magnetic torque.

2. Buried or Interior Type

The alternative to mounting magnets on the surface of the rotor is to embed them in the interior of the rotor as shown in the below figure. Interior magnet designs offer the advantage of mechanical robustness and a smaller airgap. The reluctance torque obtained is much higher than that in the insert magnet motors.

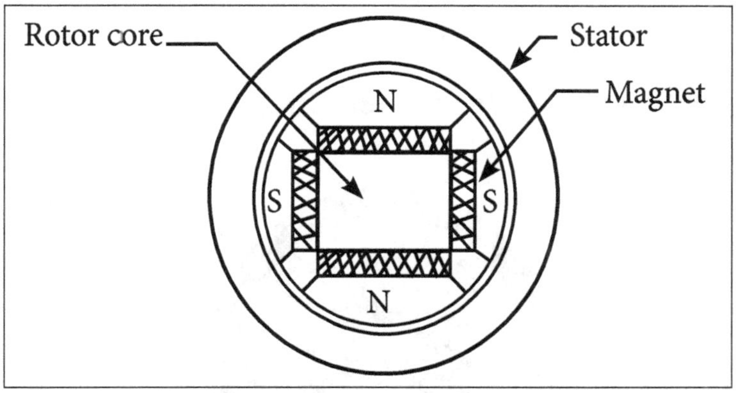

Buried or interior type PMSM.

Mechanical Power Developed in the Motor

The expression for power and torque of projecting type Air face magnet is same as that of salient pole wound field motor. Interior of insert type motor is similar to that of cylindrical rotor wound field type.

The field excitation is provided by the permanent magnet and the rotor rotates continuously and attains synchronous speed at rated frequency.

The mechanical power developed in the motor is given by:

$$P_m = 3EI_s \cos\left(s' - \frac{\pi}{z}\right)$$

$$P_m = 3\,X_s\,I_s\,I_f\,\sin s$$

$$K = \frac{3X_s}{\omega_s} = \text{constant}\,T = \frac{P_m}{\omega_s} = k\,l_s\,l_f\,\sin s$$

$$T_s = \pm\,kI_fI_s$$

Hence, torque is directly proportional to I_S.

Equivalent Circuit

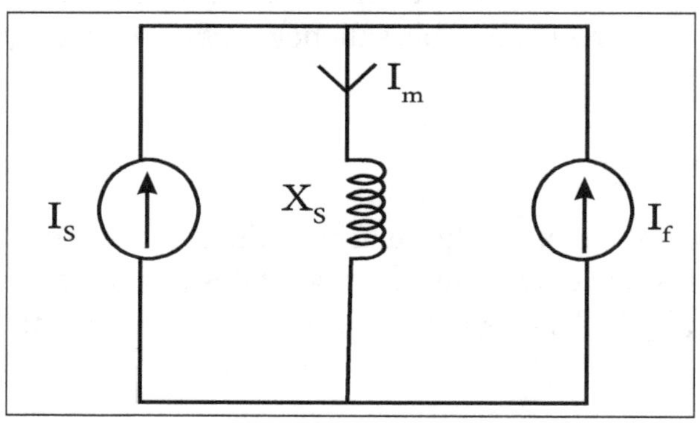

Equivalent Circuit.

Construction and Working of Permanent Magnet Synchronous Motor

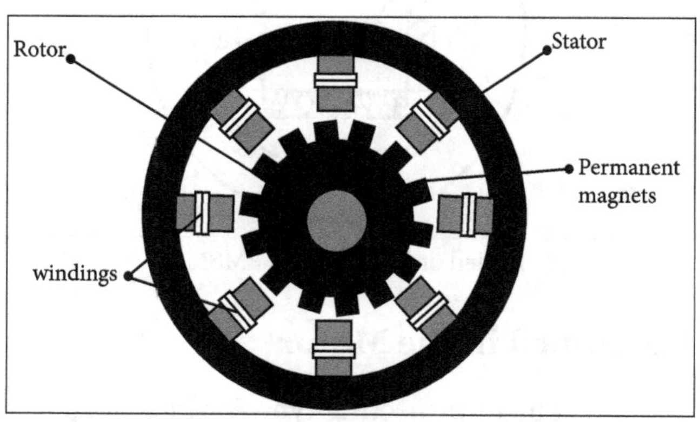

Permanent Magnet Synchronous Motor.

The armature winding and shape of permanent magnet are designed so that the flux density distribution of air gap is sinusoidal.

Stator: It consists of armature winding. A sinusoidal PMAC motor has distributed armature winding in stator. The thickness of stator lamination depends upon the frequency of armature source voltage. The machine wind age losses can be reduced by having longer air gaps.

Rotor: Rotor is made up of permanent magnets. Rare earth (cobalt- samarium) magnets are employed. The following are the 4 types of rotor geometries.

Peripheral Type Rotor

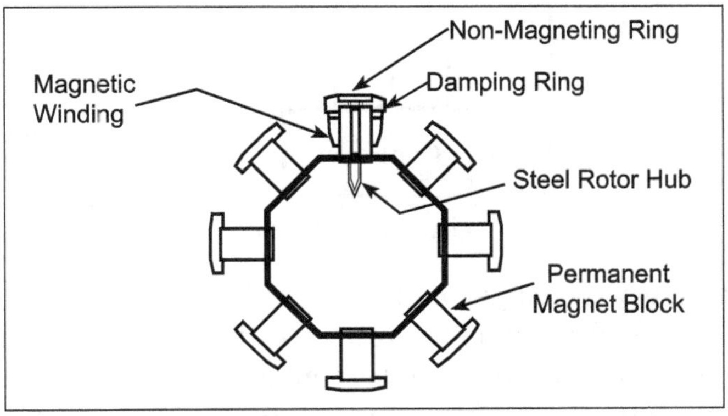

Peripheral Type Rotor.

In peripheral type rotor, the permanent magnets are located on the rotor periphery. The flux pattern of permanent magnet is radial.

Claw-Pole Type Rotor

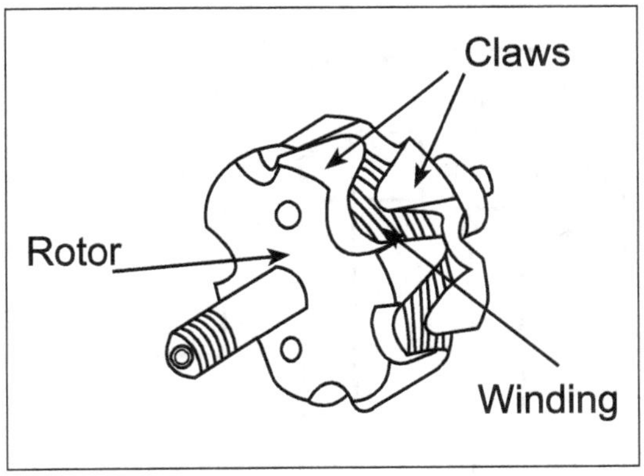

Claw-Pole Type Rotor.

The permanent magnets are disc-shaped and magnetized axially. The long soft-extensions of the construction come out axially from the periphery of the discs like claws (Lundell Poles).

Principle of Operation

When a three-phase AC supply is given to the stator of permanent magnet synchronous motor, it produces rotating magnetic field. So the stator flux is produced. Due to the presence of permanent magnet in the rotor, the rotor flux is produced.

Due to the interaction of the stator flux and the rotor flux, the torque is generated. To get the maximum generated torque, the angle between the stator flux and the rotor flux is kept at 90°.

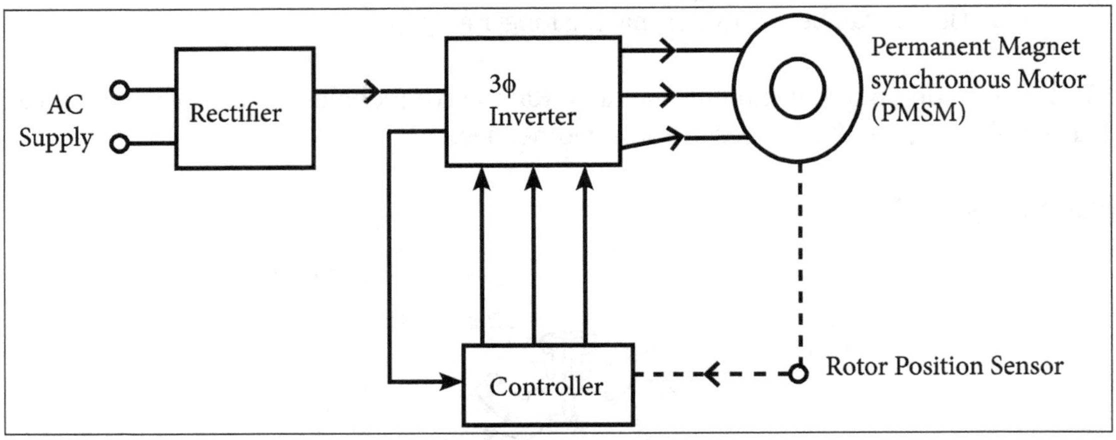

Operation of PMSM.

The armature supply frequency is changed in proportion to the rotor speed changes. So that stator field always moves at same speed as rotor. The rotor position sensor is required for accurate tracking of speed in order to prevent motor from pulling out of step and to avoid instability. The PMSM drive is used in high performance servo drives in-spite of its high cost.

Closed Loop Control of PMSM Motor

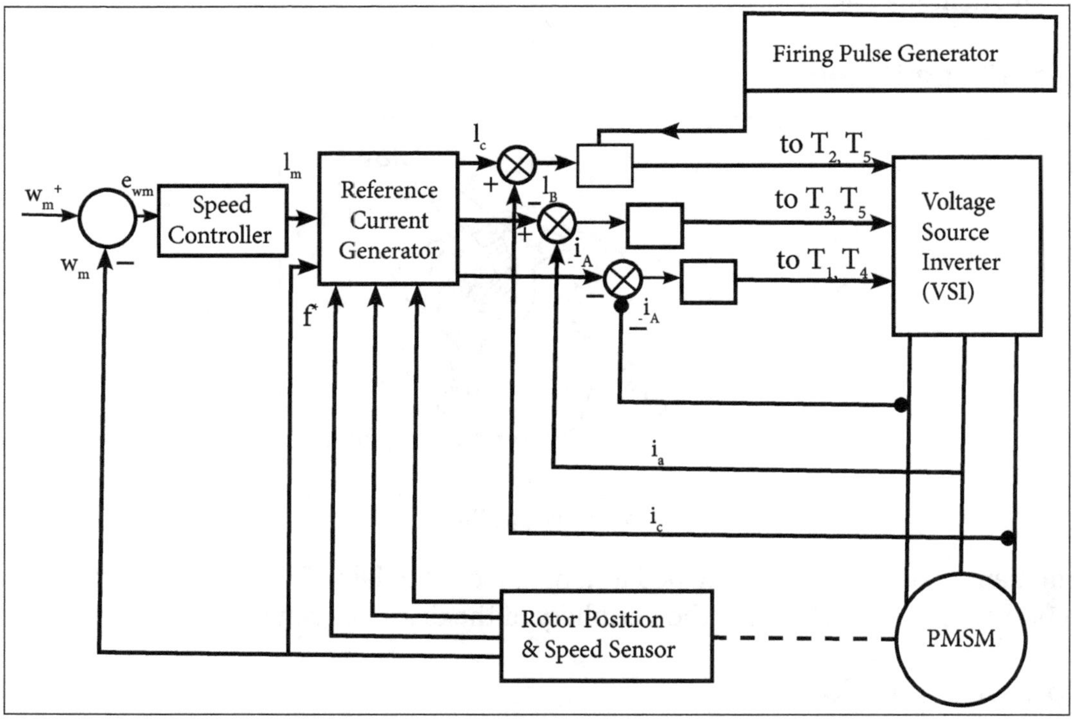

(a) Closed loop position control of Permanent synchronous motor (PMSM) drive.

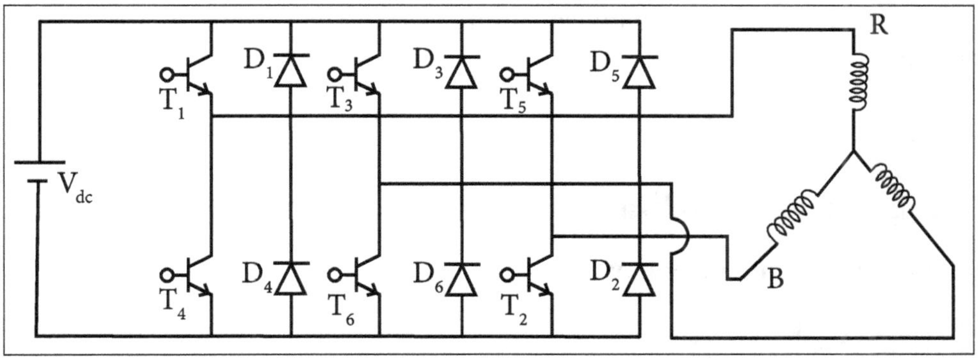

(b) VSI fed PMSM.

Figure (a) shows the closed loop variable speed drive employing PMSM fed from current regulated VSI. Based on the current amplitude command I_m* and frequency command f*, reference current generator generates sinusoidal reference currents i^*_A, i^*_B and i^*_C. These references currents are compared with respective motor currents i_A, i_B and i_C in comparators with hysteresis to generate base drives for switches.

The actual speed of PMSM (w_m) sensed by speed sensor is compared with reference speed. The error in speed is processed through speed controller. The output of speed controller sets a reference for amplitude and polarity of stator current I_s*. Rotor position sensors generate the stator current templates for the three phases.

If Speed Error = +ve, machine will work as motor and drive will accelerate to reference speed ω_m*.

If Speed Error = -ve, braking will decelerate motor to reference speed ω_m*.

The rotating fields produced by the stator and rotor should have an angle of 90° for producing the maximum torque for a given stator current. The stator field will be along phase R axis when current in phase R reaches its positive peak.

The rotor South Pole axis, at this instant must be 90° electrical behind. Therefore, rotor south pole axis must be 180° electrically behind phase R axis at the positive zero crossing of phase R current. This information is used to locate the rotor position.

Advantages

- Low torque ripple.
- Excellent dynamic performance.

Drawback

An expensive rotor position sensor or resolver is required to generate the sinusoidal current template based on the rotor position.

Permanent Magnet Machines

Permanent Magnet machines are one state variable simpler than their wound-field counterparts.

They may be accurately viewed as having constant field current. Assuming that we can define the internal (field) flux as:

$$\psi_o = x_{ad} i_{fo} \quad \text{...(1)}$$

Model: Voltage Driven Machine

In the case of a voltage driven machine:

$$\begin{bmatrix} i_d \\ i_{kd} \end{bmatrix} = \begin{bmatrix} x_d & x_{ad} \\ x_{ad} & x_{kd} \end{bmatrix}^{-1} \begin{bmatrix} \psi_d - \psi_o \\ \psi_{kd} - \psi_o \end{bmatrix} \quad \text{...(2)}$$

$$\begin{bmatrix} i_d \\ i_{kq} \end{bmatrix} = \begin{bmatrix} x_q & x_{aq} \\ x_{aq} & x_{kq} \end{bmatrix}^{-1} \begin{bmatrix} \psi_q \\ \psi_{kq} \end{bmatrix} \quad \text{...(3)}$$

The simulation model then has six states such as:

$$\frac{d\psi_d}{dt} = \omega_o \tilde{o}_d + \omega\psi_q - \omega_o r_a i_d$$

$$\frac{d\psi_q}{dt} = \omega_o \tilde{o}_q - \omega\psi_d - \omega_o r_a i_q$$

$$\frac{d\psi_{kd}}{dt} = -\omega_o r_{kd} i_{kd}$$

$$\frac{d\psi_{kq}}{dt} = -\omega_o r_{kq} i_{kq}$$

$$\frac{d\omega}{dt} = \frac{\omega_o}{2H}\left(\psi_d i_q - \psi_d i_q + T_m\right)$$

$$\frac{d\delta}{dt} = \omega - \omega_o$$

Current-Driven Machine Model

In the case of a current-driven machine, rotor currents required in the simulation are:

$$i_{kd} = \frac{1}{x_{kd}}\left(\psi_{kd} - x_{ad} i_d - \psi_o\right)$$

$$i_{kq} = \frac{1}{x_{kq}}\left(\psi_{kq} - x_{aq} i_q\right)$$

Here, the "flux behind sub transient reactance" is:

$$\psi''_d = \frac{x_{kdl}\Psi_o + x_{ad}\Psi_{kd}}{x_{ad} + x_{kdl}}$$

and the sub-transient reactance is given by:

$$x''_d = x_{al} + x_{ad} \| x_{kdl}$$

On the quadrature axis:

$$\psi''_q = \frac{x_{ad}\Psi_{kq}}{x_{ad} + x_{kql}}$$

And

$$x''_q = x_{al} + x_{aq} \| x_{kql}$$

In this case, there are only four state equations.

$$\frac{d\psi_{kd}}{dt} = -\omega_o r_{kd} i_{kd}$$

$$\frac{d\psi_{kq}}{dt} = -\omega_o r_{kq} i_{kq}$$

$$\frac{d\omega}{dt} = \frac{\omega_o}{2H}\left(e''_q i_q + e''_d i_d + T_m\right)$$

$$\frac{d\delta}{dt} = \omega - \omega_o$$

The interconnections to and from the network are same as in the case of a wound-field machine. In the "forward" direction, from network to machine:

$$i_d = i_a \cos(\theta) + i_b \cos\left(\theta - \frac{2\pi}{3}\right) + i_c \cos\left(\theta + \frac{2\pi}{3}\right)$$
$$i_q = -i_a \sin(\theta) - i_b \sin\left(\theta - \frac{2\pi}{3}\right) - i_c \sin\left(\theta + \frac{2\pi}{3}\right)$$

and in the reverse direction, from the machine to the network.

$$e''_a = -\frac{}{\omega_o}\left(e''_q \sin(\theta) - e''_d \cos(\theta)\right) + \frac{1}{\omega_o}\cos(\theta)\frac{de''}{dt} + \frac{1}{\omega_o}\sin(\theta)\frac{de''}{dt}$$

Problems

1. A 3-phase 400V, 50Hz, 6 pole, star connected round rotor synchronous motor has $Z_g = 0 + j2\ \Omega$. Load torque, proportional to speed squared is 340 Nm at rated synchronous speed. The speed of the motor is lowered by keeping v/f constant and maintaining unity PF by field control of the motor. For the motor operation at 600 rpm, let us calculate (i) supply voltage, (ii) the armature current, (iii) excitation voltage, (iv) load angle and (v) pullout torque. Neglect rotational losses.

Solution:

Given:

3-phase 400V, 50Hz, 6 pole motor

$Z_g = 0 + j2\ \Omega.$

Speed = 340Nm

Formula used:

$$X_s = \frac{f}{f_m} \times (X)$$

$$E = \sqrt{V_S^2 + (I_S X_S)^2}$$

Armature Current (I_a):

$$f = \frac{PN}{120} = \frac{6 \times 600}{120} = 30\,\text{Hz}[\because N = 600\,\text{rpm}]$$

$$\frac{V_s}{30} = \frac{400}{50} \Rightarrow V_s = 240\,\text{V}$$

$$N_1 = \frac{120f}{P} = \frac{120 \times 50}{6} = 1000\,\text{rpm}$$

$$T_L = T_r \left(\frac{N}{N_1}\right)^2 = 340\left(\frac{600}{1000}\right)^2 = 122.4\,\text{Nm}$$

$$P = T_L \omega = 122.4 \times \frac{2\pi N}{60} = 7690.62\,\text{W}$$

$$P = \sqrt{3}\ \text{Vs IA} \cos\phi\ [\because\ \cos\phi = 1 = \text{unity power factor}]$$

$$I_A = 18.5\,\text{A}$$

Excitation Voltage (E)

$$Z = (0 + 2j)\Omega,\ f = 30\ \text{Hz},\ f_m = 50\text{Hz}\ [\because X = 2]$$

$$X_s = \frac{f}{f_m} \times (X)$$

$$= \frac{30}{50} \times 2 = 1.2\Omega$$

$$V_S = \frac{240}{\sqrt{3}} = 138.57\,V,\ I_A = 18.5A$$

$$E = \sqrt{V_S^2 + (I_S X_S)^2} = 140.33V/\text{phase}$$

$$E_l = E \times \sqrt{3} = 243.1\,V.$$

Load Angle (δ)	Pull Out Torque (T_P)
$\delta = \tan^{-1}\left(\frac{I_S X_S}{V_S}\right) = 9.1^{o}$	$T_P = \frac{3EV}{\omega_S V_S} = \frac{3EV(60)}{2\pi N V_s}$ $T_P = 773.8Nm$

2.4 Poly-phase Induction Machines: Introduction

A polyphase induction motor consists of a polyphase winding embedded in a laminated stator and a conductive squirrel cage embedded in a laminated rotor. Three phase currents flowing within the stator create a rotating magnetic field which induces a current and consequent magnetic field in the rotor.

2.4.1 Construction and Principle of Operation

The polyphase induction motor is the most commonly used industrial motor, finding application in many situations where speed regulation is not essential. It is simple and relatively inexpensive, and the absence of sliding contacts in the squirrel-cage machine reduces maintenance to a minimum.

There are two general types of polyphase induction motors such as the squirrel-cage type and the wound-rotor machine. Both the motors have an armature or stator structure similar to that of the alternating current generator, consisting of a hollow cylinder of laminated sheet steel in which there are punched longitudinal slots.

A symmetrical polyphase winding is laid in these slots which, when connected to a suitable voltage source, produces a travelling MMF wave in the air gap, rotating at a synchronous speed equal to:

$$RPM_{sync} = 120\,f/p$$

Where, f is the frequency and p the number of poles for which the stator is wound.

The squirrel-cage type of rotor is made up of sheet steel laminations keyed to the shaft and having slots punched in the periphery. The number of slots in the rotor is never a multiple of the number in the stator, thereby preventing rotor locking under light load conditions. The rotor conductors in most machines are made of aluminum alloy either molded or extruded in place in the slots, with end rings being cast as an integral part of the structure and connecting all bars at both ends.

The air-gap length between rotor and stator is kept as short as manufacturing tolerances will allow in order minimizing the magnetizing current necessary for the production of normal air-gap flux. A simple two pole, three-phase, squirrel-cage induction motor is as shown in the below figure:

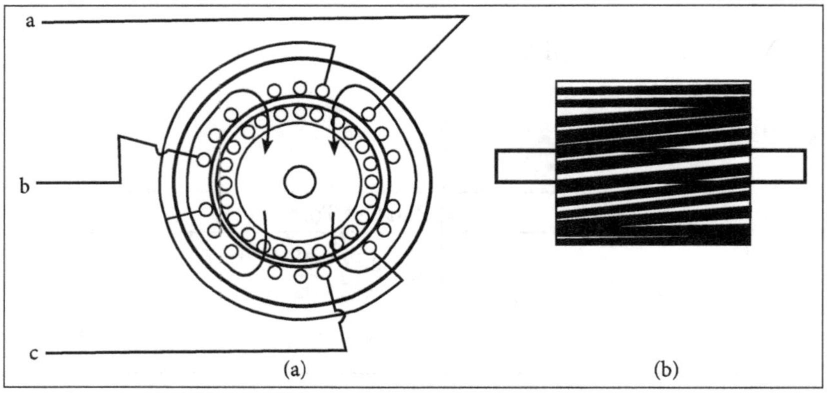

Physical construction of the squirrel-cage induction motor: (a) Cross section showing stator and rotor (b) Rotor construction.

The wound-rotor induction motor has a rotor similar to that of the squirrel-cage machine except that the short-circuited squirrel-cage winding is replaced by a three-phase insulated winding similar to that on the stator. This winding is usually wye-connected with the terminals brought out to three slip rings on the shaft. Graphite brushes connected to the slip rings provide external access to the rotor winding which is connected to a rheostatic controller, the purpose of which is to insert additional resistance in each rotor phase to improve the starting characteristics.

In practically all induction motors, either the rotor or the stator slots are skewed one slot width as shown in Figure (a). The purpose is to smooth the flux transition from one slot to the next, thereby reducing harmonics in the torque characteristic and improving the operation.

Basic Operation of the Induction Motor

The phase displacement between the voltages applied to the stator windings produces a travelling MMF or rotating magnetic field in the uniform air gap. This field links the short-circuited rotor windings and the relative motion induces short-circuit currents in them, which move about the rotor in exact synchronism with the rotating magnetic field.

It is well known that any induced current will react in opposition to the flux linkages producing it, resulting in a torque on the rotor in the direction of the rotating field. This torque causes the rotor to revolve so as to reduce the rate of change of flux linkages reducing the magnitude of the induced current and the rotor frequency.

If the rotor revolves at exactly synchronous speed, there would be no changing flux linkages about the rotor coils and no torque would be produced. However, the practical motor has friction losses requiring some electromagnetic torque, even at no load, and the system will stabilize with the rotor revolving at slightly less than synchronous speed.

The mechanical shaft load will cause the rotor to decelerate, but this increases the rotor current, automatically increasing the torque produced, and stabilizing the system at a slightly reduced speed. The difference in speed between rotor and rotating magnetic field is termed "slip" which is numerically expressed as:

$$\text{Slip} = s = \frac{(\text{synchronous speed}) - (\text{rotor speed})}{(\text{synchronous speed})}$$

This varies from a fraction of one percent at no-load to a maximum value of three or four percent under full load conditions for most properly designed machines. The speed change between no-load and full-load is small so that the squirrel-cage motor is often termed as constant-speed machine.

2.4.2 Induction Motor Equivalent Circuit

The stator windings constitute primary windings that induce flux in the rotor and stator iron. The rotor windings constitute a secondary winding that is shorted. Hence, an equivalent circuit similar to that represents the transformer is derived and appears as shown in the below figure.

Since the rotor frequency in the actual machine is dependent upon the rotor speed, all the rotor quantities must be modified to be referred to the frequency and voltage bases of the stator for inclusion in the equivalent circuit. Since the circuit represents just one phase of the actual polyphase machine, all the values are given on a per-phase basis.

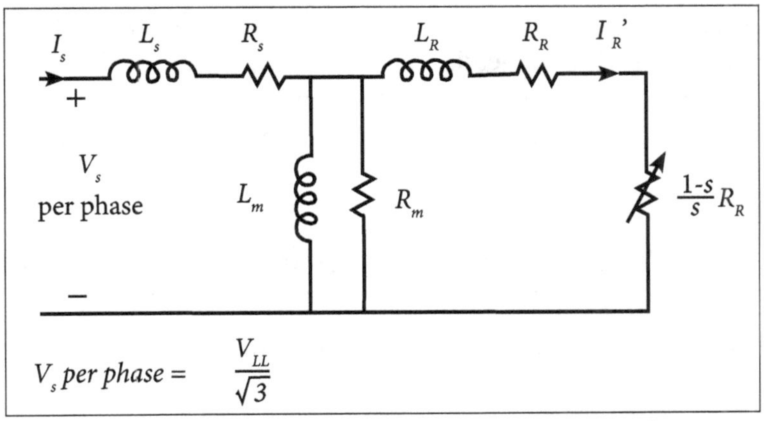

Equivalent circuit model of the induction machine per phase.

Once the equivalent circuit constants have been determined, the operating characteristics may be determined directly from it. The variable load resistance $R_R(1 - s)/s$ models the conversion of power from electrical to mechanical form.

The power absorbed by this resistance is equal to the mechanical output power of the machine P_o for a three-phase machine. This power is equal to:

$$P_o = 3\|I'_R\|^2 \frac{1-S}{S} R_R$$

Similarly, the torque is proportional to the power divided by the speed. Since the speed is proportional to 1 – s, the torque is given by:

$$T = \frac{P_o}{(1-s)\omega_s} = 3\|I'_R\|^2 \frac{R_R}{s\omega_s}$$

Where,

ω_s is the synchronous speed, in radians per second.

The torque is expressed in Newton-meters. Note that the synchronous speed in rpm is related to the applied stator frequency f. The torque expressed in foot-pounds is given by:

$$T = 3K\|I'_R\|^2 \frac{R_R}{s} \text{ foot-pounds}$$

Where K = 0.058 p/f.

The losses may be evaluated by realizing that R_s and R_R represent the stator and rotor resistances per phase respectively and R_m models the core loss.

For the usual constant speed application, the mechanical wind age (i.e., the resistance of air to rotation of the shaft) and bearing friction losses are constant. Then R_m can also model these losses and the total of these losses is called the stray power loss.

The inductance L_m models the magnetization characteristic of the complete flux path. This is dominated by the characteristic of the air gap between stator and rotor. A significant difference between the numerical values of the parameters of the induction machine versus the transformer is the relatively low value of L_m. This low L_m leads to a substantial magnetizing current that is typically similar in magnitude to the current in the effective load resistance $R_R(1 - s)/s$ at full load. In consequence, induction motors exhibit relatively low power factors, especially at light load.

2.5 Steady-state Performance Equations of the Induction Motor

The steady state performance of a polyphase induction motor can be obtained using per phase equivalent circuit.

V_1 = The input voltage per phase.

R_1, X_1 = The resistance and leakage reactance of the stator per phase.

R_2, X_2 = The resistance and leakage reactance of the rotor per phase.

X_m = The magnetizing reactance.

R_c = The core loss resistance.

I_1, I_2 = Stator and rotor currents per phase.

The developed torque is given as:

$$T = 3(I_2)^2 R_2/(s\omega_s) N-m$$

Where,

ω_r = Rotor speed in radians/sec.

ω_s = Synchronous speed in radians/sec.

s = Slip

Its parameters could be determined by no load and blocked rotor test, where the former determines.

R_c and X_m while the latter yields R_1, R_2, X_1, X_2.

$$X_1 = R_1 + jX_1$$

$$X_2 = R_2/S + jX_2$$

$$S = \omega_s - \omega_r/\omega$$

The torque $T = 3(I_2)^2 R_2/s$

Input power $= 3V_1 * I_1 * pf$

Output power $= (1-S)\ \omega_s$

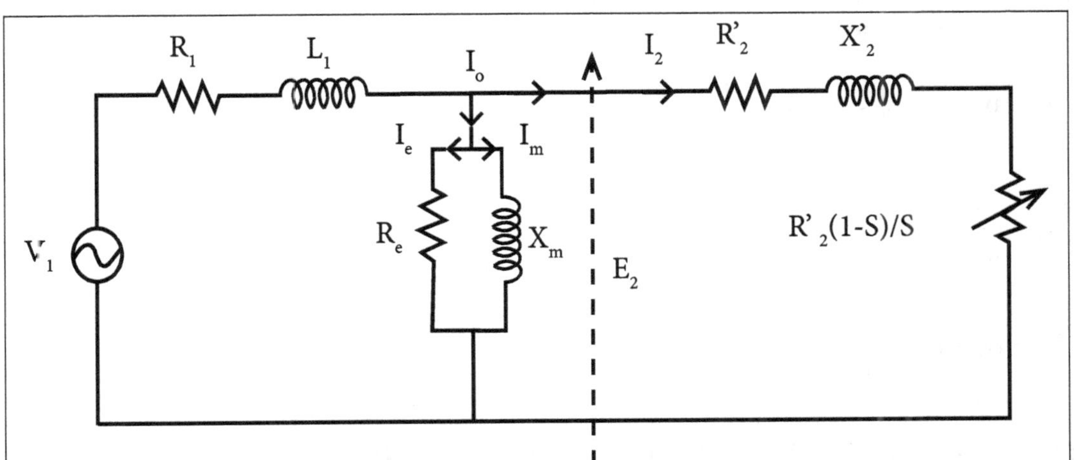

Per phase equivalent circuit of the polyphase induction motor.

Start the motor by applying normal frequency, reduce the voltage to the stator and gradually increase the voltage to its rated value under no load condition. For slip ring motor, short circuit the slip rings before starting. Note down the readings of the ammeter, voltmeter, wattmeter and tachometer at different voltages.

2.5.1 Measurement of Motor Parameters

The equivalent circuit constants may be determined in the same manner as those of the transformer. If the shaft coupling is disconnected, the power output will be zero and the load resistance $R_R(1\text{-}s)/s$ approaches infinity.

For all the practical purposes, the series constants may be neglected and the shunt constants obtained by measuring the voltage, current and power under these conditions where:

$$Z_m = \frac{V}{\sqrt{3}I}$$

$$R_m = \frac{V^2}{P}$$

$$L_m = \frac{1}{\omega\sqrt{\left(\frac{I}{Z_m}\right)^2 - \left(\frac{I}{R_m}\right)^2}}$$

Where,

I = Line current.

P = Total three-phase power.

V = Line-to-line voltage.

If the rotor is blocked so as to prevent rotation and a balanced low-voltage three phase source connected to the stator terminals, the load resistance $R_R(1 - s)/s$ will reduce to zero and the shunt branch may be neglected. Then:

$$R_e = R_R + R_s \text{ per phase} = \frac{P}{3I^2}$$

$$Z_e = \frac{V}{\sqrt{3}I}$$

$$L_e = L_R + L_s = \frac{1}{\omega}\sqrt{Z_e^2 - R_e^2}$$

R_s Per phase may be determined by passing direct current through any two terminals of the stator, recording the voltage drop and dividing the resultant resistance by two. Then:

$$R_R = R_e - R_s$$

It is usually accurate to assume equal stator and rotor leakage inductances, so that $L_s = L_R = L_e/2$.

2.5.2 Dynamic Modeling of Induction Machines

Dynamic equations of the induction motor are expressed in dq frame. The idea of a revolving reference frame, DQ, is introduced to transform the AC components of the vectors in the stator frame into DC signals and formulas for the straight and inverse abc→dq and dq→DQ transformations are provided.

Space Vectors of Motor Variables

Space vectors of three-phase variables, such as the voltage current, or flux, are very convenient for the analysis and control of induction motors. The vector of total stator MMF, $\vec{\mathscr{F}}_s$ is a vectorial sum of phase MMFs, $\vec{\mathscr{F}}_{as}, \vec{\mathscr{F}}_{bs}$ and $\vec{\mathscr{F}}_{cs}$. i.e:

$$\vec{\mathscr{F}}_s = \vec{\mathscr{F}}_{as} + \vec{\mathscr{F}}_{bs} + \vec{\mathscr{F}}_{cs} = \vec{\mathscr{F}}_{as} + \vec{\mathscr{F}}_{bs} e^{j\frac{2}{3}\pi} + \vec{\mathscr{F}}_{cs} e^{j\frac{4}{3}\pi} \quad \text{...(1)}$$

Where,

$\vec{\mathscr{F}}_{as}, \vec{\mathscr{F}}_{bs}$ and $\vec{\mathscr{F}}_{cs}$ denotes the magnitudes of $\vec{\mathscr{F}}_{as}, \vec{\mathscr{F}}_{bs}$ and $\vec{\mathscr{F}}_{cs}$ respectively.

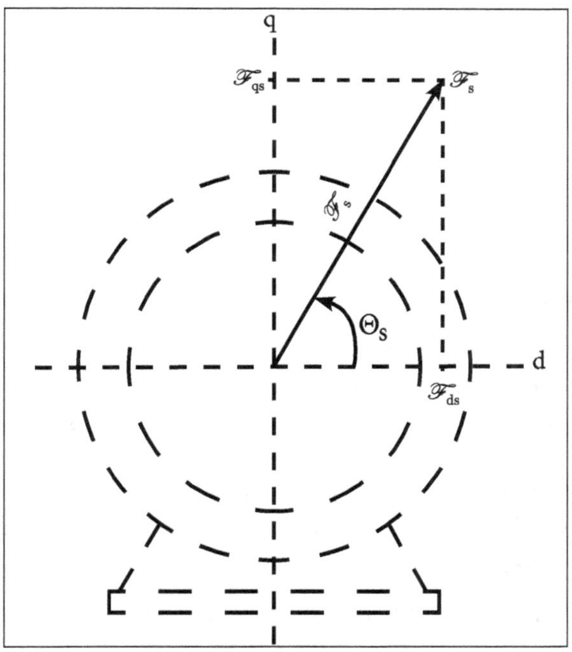

Space vector of stator MMF.

In the stationary set of stator coordinates, dq, the vector of stator MMF can be expressed as a complex variable, $\vec{\mathscr{F}}_s = \vec{\mathscr{F}}_{ds} + j\vec{\mathscr{F}}_{qs} = \vec{\mathscr{F}}_s e^{j\Theta s}$ as depicted in the figure. Because:

$$e^{j\frac{2}{3}\pi} = -\frac{1}{2} + j\frac{\sqrt{3}}{2} \quad \text{...(2)}$$

And

$$e^{j\frac{4}{3}\pi} = -\frac{1}{2} - j\frac{\sqrt{3}}{2} \quad \text{...(3)}$$

Equation (1) can be rewritten as:

$$\mathscr{F}_s = \mathscr{F}_{ds} + j\mathscr{F}_{qs} = \mathscr{F}_{as} - \frac{1}{2}\mathscr{F}_{bs} - \frac{1}{2}\mathscr{F}_{cs} + j\left(\frac{\sqrt{3}}{2}\mathscr{F}_{bs} - \frac{\sqrt{3}}{2}\mathscr{F}_{cs}\right) \quad ...(4)$$

Which explains the $abc \to dq$ transformation. For the stator MMFs,

$$\begin{bmatrix} \mathscr{F}_{ds} \\ \mathscr{F}_{qs} \end{bmatrix} = \begin{bmatrix} 1 & -\frac{1}{2} & -\frac{1}{2} \\ 0 & \frac{\sqrt{3}}{2} & -\frac{\sqrt{3}}{2} \end{bmatrix} \begin{bmatrix} \mathscr{F}_{as} \\ \mathscr{F}_{bs} \\ \mathscr{F}_{cs} \end{bmatrix} \quad ...(5)$$

And

$$\begin{bmatrix} \mathscr{F}_{as} \\ \mathscr{F}_{bs} \\ \mathscr{F}_{cs} \end{bmatrix} = \begin{bmatrix} \frac{2}{3} & 0 \\ -\frac{1}{3} & \frac{1}{\sqrt{3}} \\ -\frac{1}{3} & -\frac{1}{\sqrt{3}} \end{bmatrix} \begin{bmatrix} \mathscr{F}_{ds} \\ \mathscr{F}_{qs} \end{bmatrix} \quad ...(6)$$

Transformation equations (5) and (6) apply to all the three-phase variables of the induction motor, which add up to zero.

Stator MMFs are true vectors, because their direction and polarity in the real space of the motor can be easily ascertained. Because, an MMF is a product of the current in a coil and the number of turns of the coil, the stator current vector, i_s, can be obtained by dividing $\mathscr{F}_s$ by the number of turns in the phase of the stator winding. This is tantamount to apply the $abc \to dq$ transformation to currents, i_{as}, i_{bs} and i_{cs} in the individual phase windings of the stator.

The stator voltage vector v_s is obtained using the same transformation to stator phase voltages, v_{as}, v_{bs} and v_{cs}. It can be argued to which extent i_s and v_s are true vectors, but from the viewpoint of analysis and control of induction motors, this issue is irrelevant. It must be mentioned that the $abc \to dq$ and $dq \to abc$ transformation matrices in (5) and (6) are not the only ones encountered in the literature.

When the stator phase MMFs are balanced, the magnitude $\mathscr{F}_s$ of the space vector $\mathscr{F}_s$ of the stator MMF is 1.5 times higher than the magnitude (peak value), $\mathscr{F}_{as}$ of phase MMFs. This coefficient applies to all other space vectors. Then, the vector magnitude equals the peak value of the corresponding phase quantities. On the other hand, if the product of magnitudes, v_s and i_s is equal to the apparent power supplied to the stator, the matrices in (5) and (6) should be multiplied by √(2/3) and √(3/2), respectively.

In practical ASDs, the voltage feedback, if needed, is usually obtained from the voltage sensor, which placed at the DC input to the inverter, measures the DC-link voltage, v_i. The line-to-line and line-to-neutral stator voltages are determined on the basis of the current values, a, b and c of the switching variables of the inverter.

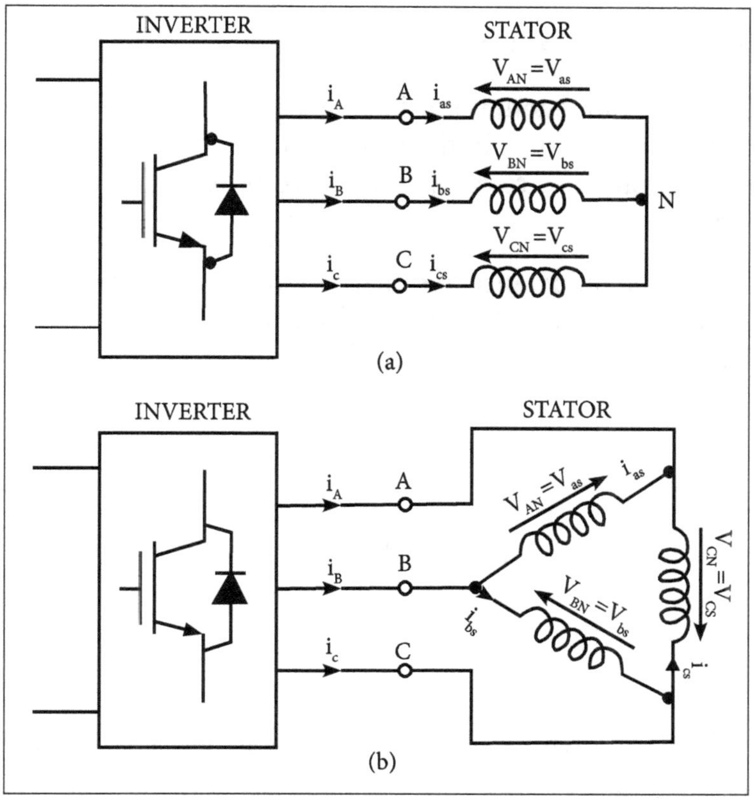

Stator currents and voltages: (a) Wye-connected stator and (b) Delta-connected stator.

Depending on whether the phase windings of the stator are connected in delta or wye, the stator voltages v_{as}, v_{bs} and v_{cs} constitute the respective line-to-line or line-to-neutral voltages. Specifically, in a delta-connected stator, $v_{as} = v_{AB}, v_{bs} = v_{BC}$ and $v_{cs} = v_{CA}$, while in a wye-connection, $v_{as} = v_{AN}$, $v_{bs} = v_{BN}$ and $v_{cs} = v_{CN}$.

The current feedback is typically provided by two current sensors in the output lines of the inverter. The sensors measure the currents i_A and i_C, and if the stator is connected in wye, its phase currents are easily determined as $i_{as} = i_A, i_{bs} = -i_A - i_C$ and $i_{cs} = i_C$.

Because of the symmetry of all the three phases of the motor and symmetry of the control of all the phases of the inverter, the phase stator currents in a delta-connected motor can be assumed to add upto zero. Consequently, they can be found as $i_{as} = (2\,i_A + i_C)/3$, $i_{bs} = (-i_A - 2i_C)/3$ and $i_{cs} = (-i_A + i_C)/3$.

Voltages and currents in the wye-connected and delta-connected stators are shown in the above figure. In addition to the space vectors of the stator voltage, v_s and current, i_s, four other three-phase variables of the induction motor will be expressed as space vectors. These are the rotor current vector, i_r and three flux-linkage vectors, commonly called as flux vectors, are stator flux vector, λ_s, air-gap flux vector, λ_m and rotor flux vector, λ_r.

The air-gap flux is smaller than the stator flux by small amount of leakage flux in the stator and similarly, the rotor flux is only slightly reduced with respect to the air-gap flux, due to flux leakage in the rotor.

Dynamic Equations of the Induction Motor

The dynamic T-model of the induction motor in the stator reference frame, with motor variables expressed in the vector form, is shown in the below figure. Symbol p denotes the differentiation operator, d/dt, while L_{ls}, L_{lr} and L_m are the stator and rotor leakage inductances and the magnetizing inductance, respectively.

The sum of the stator leakage inductance and magnetizing inductance is called the stator inductance and denoted by L_s. Analogously, the rotor inductance, Lr is defined as the sum of the rotor leakage inductance and magnetizing inductance. Thus,

$$L_s = L_{ls} + L_m \text{ and } L_r = L_{lr} + L_m$$

Where,

$$L_s = X_s/\omega$$

$$L_r = X_r/\omega$$

The dynamic model allows the derivation of the voltage-current equation of the induction motor. Using space vectors, the equation can be written as:

$$\frac{d_i}{dt} = Av + Bi \qquad \text{...(7)}$$

Where,

$$i = \left[i_{ds}\ i_{qs} i_{dr}\ i_{qr}\right]^T \qquad \text{...(8)}$$

$$v = \left[v_{ds}\ v_{qs} v_{dr}\ v_{qr}\right]^T \qquad \text{...(9)}$$

$$A = \frac{1}{L_\sigma^2}\begin{bmatrix} L_r & 0 & -L_m & 0 \\ 0 & L_r & 0 & -L_m \\ -L_m & 0 & L_s & 0 \\ 0 & -L_m & 0 & L_s \end{bmatrix} \qquad \text{...(10)}$$

$$B = B(\omega_o) = \frac{1}{L_\sigma^2}\begin{bmatrix} -R_sL_r & \omega_oL_m^2 & R_rL_m & \omega_oL_rL_m \\ -\omega_oL_m^2 & -R_sL_r & -\omega_oL_rL_m & R_rL_m \\ R_sL_m & -\omega_oL_sL_m & -R_rL_s & -\omega_oL_sL_r \\ \omega_oL_sL_m & R_sL_m & \omega_oL_sL_r & -R_rL_s \end{bmatrix} \qquad \text{...(11)}$$

$$L^2 = L_s L_r - L_m^2 \qquad ...(12)$$

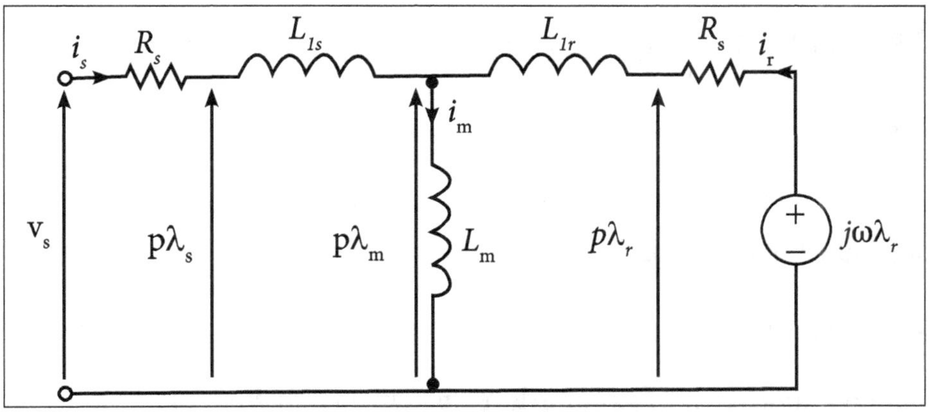

Dynamic T-model of the induction motor.

Symbols i_{dr} and i_{qr} in (9) denote the components of the rotor current vector, i_r. In the squirrel-cage motor, the corresponding components, v_{dr} and v_{qr} of the rotor voltage vector, v_r are both zero because the rotor windings are shorted. The stator and rotor fluxes are related to the stator and rotor current as,

$$\begin{bmatrix} \lambda_s \\ \lambda_r \end{bmatrix} = \begin{bmatrix} L_s & L_m \\ L_m & L_r \end{bmatrix} \begin{bmatrix} i_s \\ i_r \end{bmatrix} \qquad ...(13)$$

The stator flux can also be obtained from the stator voltage and current as:

$$\frac{d\lambda_s}{dt} = v_s - R_s i_s \qquad ...(14)$$

Or

$$\lambda_s = \int_0^t (v_s - R_s i_s) dt + \lambda_s(0) \qquad ...(15)$$

While the rotor fluxes in the squirrel-cage motor satisfies the equation:

$$\frac{d\lambda_r}{dt} = j\omega_o \lambda_r - R_r i_r \qquad ...(16)$$

Finally, the developed torque can be expressed in several forms, such as:

$$T_M = \frac{2}{3} p_p \mathrm{Im}\{i_s \lambda_s^*\} = \frac{2}{3} P_p \left(i_{qs\lambda ds} - i_{ds}\lambda_{qs}\right) \qquad ...(17)$$

$$T_M = \frac{2}{3} P_p \frac{L_m}{L_r} \mathrm{Im}\{i_s \lambda_T^*\} = \frac{2}{3} P_p \frac{L_m}{L_r} \left(i_{qs}\lambda_{qr} - i_{ds}\lambda_{qr}\right) \qquad ...(18)$$

Or

$$T_M = \frac{2}{3}P_p L_m \, \mathrm{Im}\{i_s i_r^*\} = \frac{2}{3}P_p Lm\left(i_{qs}i_{dr} - i_{ds}i_{qr}\right) \qquad ...(19)$$

Where* denotes a conjugate vector.

The rather abstract term $\mathrm{Im}\left({}_{is}\lambda_s{}^*\right)$ in (17) and the analogous terms in equation (18) and (19) represent a vector product of the involved space vectors. For instance,

$$\mathrm{Im}\left(i_s\lambda_s^*\right) = i_s\lambda_s \sin\left[\angle\left(i_s, \lambda_s\right)\right] \qquad ...(20)$$

The above equation implies that the torque developed in an induction motor is proportional to the product of magnitudes of space vectors of two selected motor variables and the sine of angle between these two vectors. It can be seen that all the torque equations are non-linear, as each of them includes a difference of products of two motor variables. Equation (7) is non-linear, too, because of the variable ω_0 appearing in matrix B.

Revolving Reference Frame

In the steady state, space vectors of motor variables revolve in the stator reference frame with the angular velocity, ω imposed by the supply source(inverter). It must be stressed that this velocity does not depends on the number of poles of stator, which indicates the abstract quality of the vectors.

Under transient operating conditions, instantaneous speed of the space vector may vary and they are not necessarily the same for all vectors, but the vectors keep revolving nevertheless. Consequently, their d and q components are AC variables, which are less convenient to analyze and utilize in a control system than the DC signals commonly used in control theory. Therefore, in addition to the static $abc \rightarrow dq$ and $dq \rightarrow abc$ transformations, the dynamic $dq \rightarrow DQ$ and $DQ \rightarrow dq$, transformations from the stator reference frame to a revolving frame and vice versa are often employed.

Usually, the revolving reference frame is selected so that it moves in synchronism with a selected space vector. The revolving reference frame, DQ, rotating with the frequency ω_e, is shown in the below figure with the stator reference frame in the background. The stator voltage vector, vs, revolves in the stator frame with the angular velocity of ω remaining stationary in the revolving frame if $\omega_e = \omega$.

Consequently, the vDS and vQS components of that vector in the latter frame are DC signals, constant in the steady state and varying in the transient states. Considering the same stator voltage vector, its $dq \rightarrow DQ$ transformation is given by:

$$\begin{bmatrix} v_{DS} \\ v_{QS} \end{bmatrix} = \begin{bmatrix} \cos(\omega_e t) & \sin(\omega_e t) \\ -\sin(\omega_e t) & \cos(\omega_e t) \end{bmatrix} \begin{bmatrix} v_{ds} \\ v_{qs} \end{bmatrix} \qquad ...(21)$$

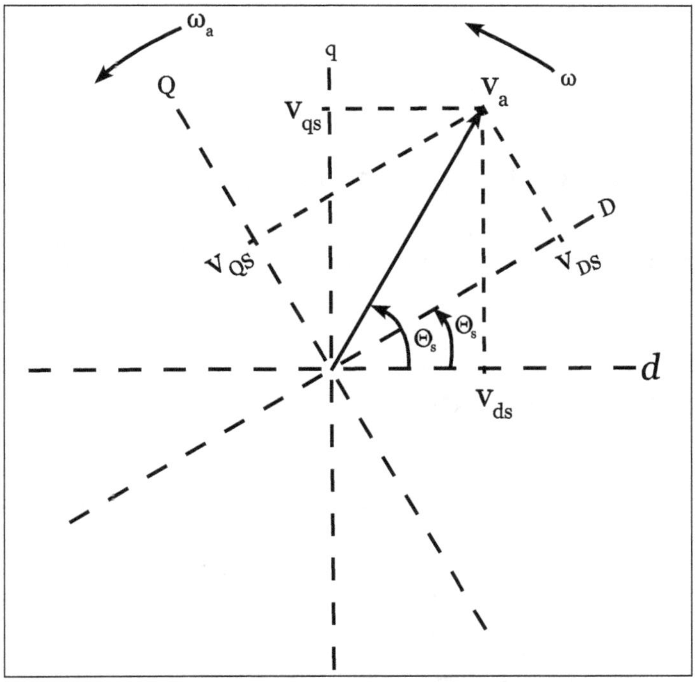

Space vector of stator voltage in the stationary and revolving reference frames.

The inverse, DQ → dq transformation is given by:

$$\begin{bmatrix} v_{ds} \\ v_{qs} \end{bmatrix} = \begin{bmatrix} \cos(\omega_e t) & -\sin(\omega_e t) \\ \sin(\omega_e t) & \cos(\omega_e t) \end{bmatrix} \begin{bmatrix} v_{DS} \\ v_{QS} \end{bmatrix} \quad ...(22)$$

3

Phase Controlled Rectifiers, Dual Converters and AC Voltage Regulators

3.1 Phase Controlled Rectifiers

Single Phase Half Wave Controlled Rectifier with Resistive Load

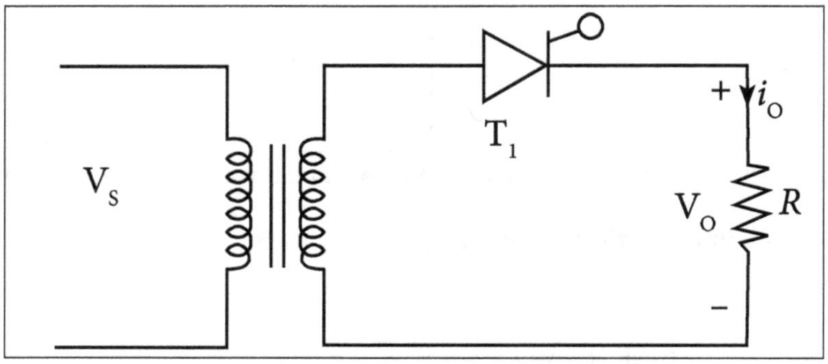

Half wave controlled rectifier with R-load.

The principle of phase controlled operation can be explained with the help of half wave controlled rectifier shown in the above figure. The secondary of the transformer is connected to resistive load through thyristor or SCR T_1.

The primary of the transformer is connected to the mains supply. In the positive cycle of the supply T_1 is forward biased. T_1 is triggered at an angle a. This is also called as triggering or firing delay angle. T_1 conducts and secondary (i.e. supply) voltage is applied to the load. Current i_0 starts flowing through the load. The output current and voltage waveforms are shown in Figure below.

Since the load is resistive, output current is given as:

$$i_o = \frac{V_o}{R}$$

Hence the shape of output current waveform is same as output voltage waveform.

At the supply, voltage drops to zero. Hence, current i_0 flowing through T_1 becomes zero and it turns off. In the negative half cycle of the supply, T_1 is reversing biased and it does not conduct. There is only one pulse of V_0 during one cycle of the supply. Hence, ripple frequency of the output voltage is given by,

$$f_{ripple} = 50\,\text{Hz i.e. supply frequency}$$

Mathematical Analysis

The average value of output voltage is given by:

$$V_{o(av)} = \frac{1}{T}\int_{0}^{T} \upsilon_o(\omega t)d\omega t$$

The period of one pulse of $v_o(\omega t)$ can be considered as $T = 2\pi$ and $v_o(\omega t) = V_m \sin \omega t$ from α to π for rest of the period $v_o(\omega t) = 0$. Hence, the above equation can be written as:

$$V_{o(av)} = \frac{1}{2\pi}\int_{\alpha}^{\pi} V_m \sin \omega t \, d\, \omega t$$

$$= \frac{V_m}{2\pi}[-\cos \omega t]_{\alpha}^{\pi}$$

$$V_{o(av)} = \frac{V_m}{2\pi}(1+\cos\alpha) \quad ...(1)$$

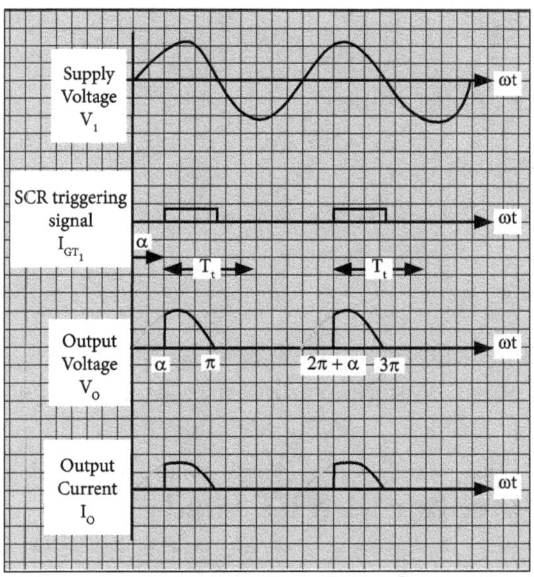

Phase controlled principle as applied to half wave controlled rectifier.

If the load resistance is R, then the average load current is given as:

$$I_{o(av)} = \frac{V_{o(av)}}{R} = \frac{V_m}{2\pi R}(1+\cos\alpha)$$

RMS value is given as:

$$V_{o(rms)} = \left[\frac{1}{T}\int_{0}^{T} \upsilon_o^2(\omega t)\, d\omega t\right]^{\frac{1}{2}}$$

$$= \left[\frac{1}{2\pi} \int_{\alpha}^{\pi} V_m^2 \sin^2 \omega t \, d\omega t \right]^{\frac{1}{2}}$$

$$= \left[\frac{V_m^2}{2\pi} \int_{\alpha}^{\pi} \frac{1 - \cos 2\omega t}{2} d\omega t \right]^{\frac{1}{2}}$$

$$= \left\{ \frac{V_m^2}{4\pi} \left[\int_{\alpha}^{\pi} d\omega t - \int_{\alpha}^{\pi} \cos 2\omega t \, d\omega t \right] \right\}^{\frac{1}{2}}$$

$$= \left\{ \frac{V_m^2}{4\pi} \left[\int_{\alpha}^{\pi} d\omega t - \int_{\alpha}^{\pi} \cos 2\omega t \, d\omega t \right] \right\}^{\frac{1}{2}}$$

$$= \frac{V_m}{2} \left[1 - \frac{\alpha}{\pi} + \frac{\sin 2\alpha}{2\pi} \right]^{\frac{1}{2}}$$

This is an expression for r.m.s output voltage.

The power transfer to load will be:

$$P_{o(av)} = \frac{V_{o(av)}^2}{R}$$

Thus, the output average voltage and power delivered by the controlled rectifier can be controlled by phase control (i.e. α). The phase control in converters means to control the delay (or triggering) angle α.

Half Wave Controlled Rectifier with RL Load

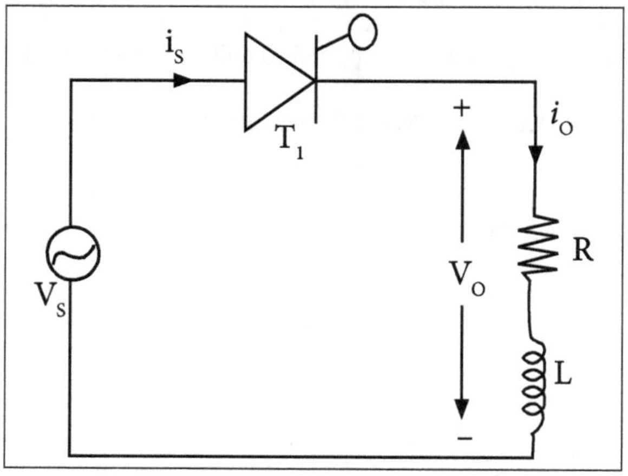

Half wave controlled rectifier with RL load.

Normally motors are inductive load. L is the armature or field coil inductance and R is the resistance of these coils. Figure above shows the circuit diagram of half wave controlled rectifier with RL load. The SCR will be forward biased in the positive half cycle of the supply. Hence SCR is applied with the firing pulses in the positive half cycle. The waveforms are as shown in the below figure:

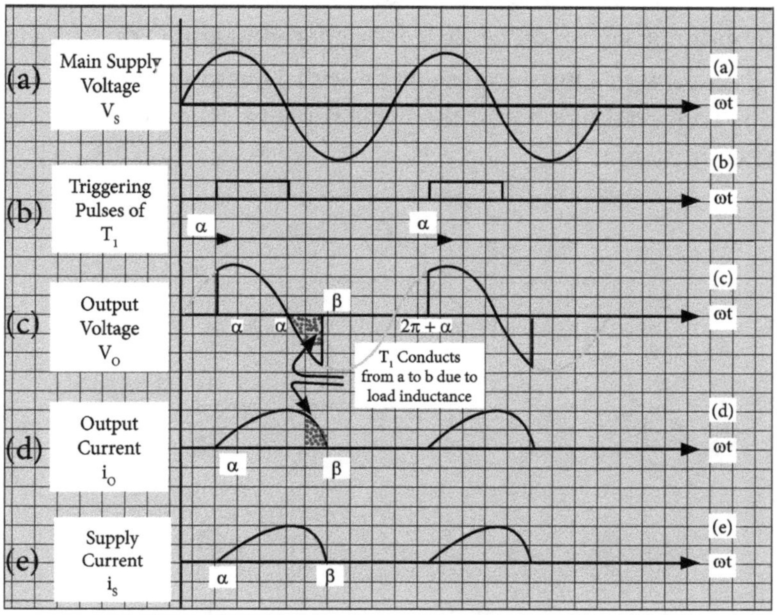

Waveforms for half wave controlled rectifier for RL load.

In Figure (a) SCR is triggered, the supply voltage appears across load. We normally neglect small voltage drop in SCR. Hence $v_o = v$, when SCR is conducting. This is shown in Figure (c). Observe that output voltage is same as supply voltage after α. Because of the RL load, output current starts increasing slowly from zero. The shape of i_o depends upon values of R and L. At π, the supply voltage becomes zero and i_o is maximum.

Due to negative supply voltage after π, SCR tries to turn-off. But energy stored in the load, inductance generates the voltage $L\frac{di_o}{dt}$. This induced voltage forward biases the SCR and d maintains it in conduction.

The basic property of inductance is that it opposes change in current. At π, the current i_o is maximum. As SCR tries to turn-off due to negative supply voltage, the output current i_o tries to go to zero. Such change in i_o is opposed by load inductance. Hence the energy stored in an inductance tries to maintain i_o.

To maintain the flow of i_o, inductance generates the voltage $L\frac{di_o}{dt}$ with polarity as shown in figure below. This voltage is higher than negative supply voltage. Hence T_1 is forward biased and it remains in conduction. The output current and supply current flow in the same loop. Hence $i_o = i_s$ all the time. The waveform of i_o is shown in above figure (d) and i_s shown in above figure (e).

After π, i_o (i.e. i_s) flows against the supply. Hence energy is consumed in the supply. i_o flows due to load inductance energy. In other words, the inductance energy is partially fed to the mains and to the load itself. Therefore, energy stored in inductance goes on reducing. Hence i_o also goes on reducing as shown in above figure (d).

At β, the energy stored in the inductance is finished. Hence i_o goes to zero. Therefore, T_1 turns-off. In the above figure(c), observe that v_o is negative from π toβ. Because T_1 conducts from π to β. Hence whenever conducts $v_o = v_s$. The SCR is triggered again at $2\pi+\alpha$. Hence output voltage remains zero from β to 2π+ α. Output current as well as supply current are also zero from to 2π+ α.At $2\pi+\alpha$, T_1 is triggered again and the cycle repeats. Here io goes to zero at Hence this is called discontinuous conduction.

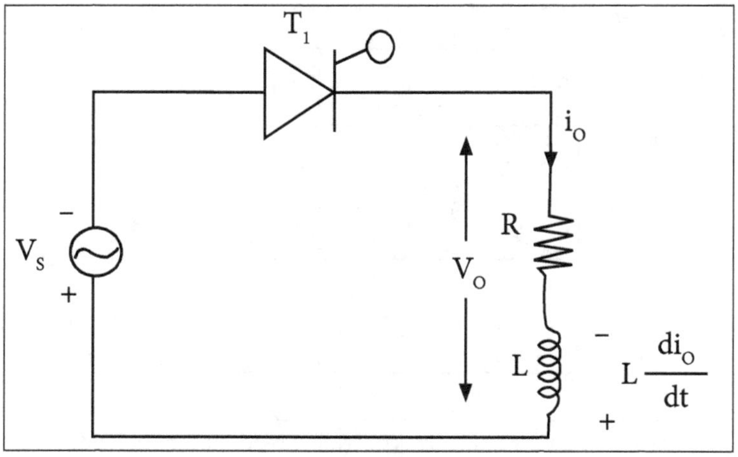

SCR conducts due to inductance voltage after π.

Half Wave Controlled Rectifier with RLE Load

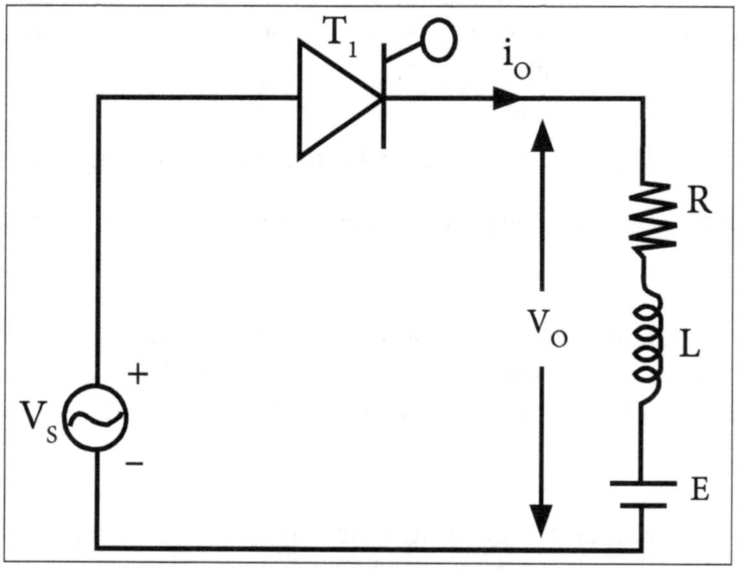

Half Wave Controlled Rectifier with RLE Load.

Figure above shows the circuit diagram of a half wave controlled rectifier with RLE load. The RLE load is normally a battery or DC motor. Normally, R and L are resistance and inductance of the motor winding and E is the back emf of the motor.

Operation and Waveforms

Mode-I

When SCR conducts, $\alpha \le \omega t \le \beta$

When SCR is conducting the supply voltage appears across the output and output voltage v_o, is same as supply voltage v_s. Output voltage waveform of below figure (b) shows that $v_o = v_s$ for $\alpha \le \omega t \le \beta$. The output current rises from zero as shown in below figure (c).

Because of inductive load the current rises from zero and achieves its peak value near π. At this time, the supply voltage reverses its polarity, and becomes negative for $\omega t > \pi$. But SCR does not turn-off due to energy stored in the load inductance. This energy generates the voltage of $L\frac{di_o(t)}{dt}$ across inductance and forward biases the SCR to keep it in conduction. Note that $L\frac{di_o(t)}{dt}$ is very large since i_ot) is not zero at π. The output current dt then goes on reducing and it becomes zero at β.

Mode-II

The SCR turns-off $\beta \le \omega t \le (2\pi + \alpha)$:

When SCR is off In the negative half cycle when SCR turns-off at a the supply voltage and back emf add and reverse bias the SCR. In the SCR voltage waveform of figure(d), observe that the supply voltage sine wave v_s is shifted towards negative side by E volts.

In other words, there is offset of -E volts. The SCR is forward biased when the supply voltage is equal to the back emf. i.e:

$$V_m \sin \omega t = E$$

or

$$\sin \omega t = \frac{E}{V_m}$$

or

$$\omega t = \sin^{-1} \frac{E}{V_m}$$

or

$$\theta_1 = \sin^{-1} \frac{E}{V_m}$$

Figure below (a) shows the location of θ_1 when supply voltage is equal to back e.m.f. Basically θ_1 and $2\pi+\theta_1$ are same points. Thus SCR cannot be triggered before θ_1. Similarly at θ_2, supply voltage is less than back e.m.f. and the SCR is reverse biased. Hence SCR cannot be triggered after after θ_2. Thus for successful triggering of SCR:

$$\theta_1 \leq \alpha \leq \theta_2$$

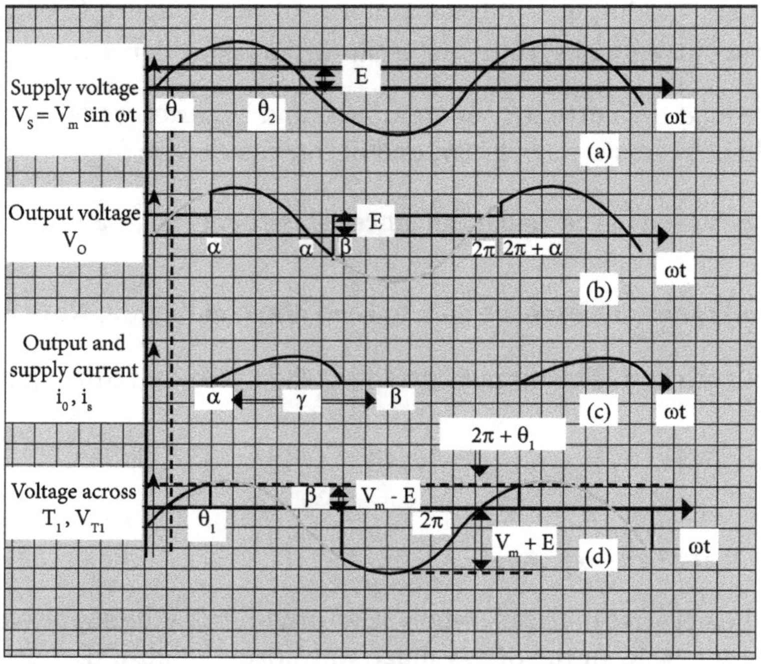

Waveforms for half wave controlled rectifier for RLE load.

Average DC Output Voltage

Assume that a sinusoidal AC input voltage is applied to a half-wave rectifier. The instantaneous value of sinusoidal AC input voltage is given by:

$$v = V\sin \omega t = V_m \sin \theta$$

Where,

V_m is the maximum voltage

$\theta = \omega t$ is the angle

The output voltage across the load can be expressed as:

$$v = V\sin \omega t = V_m \sin \theta o \leq \theta \leq \pi$$

$$= oo \leq \theta \leq 2\pi$$

$$V_{av} = V_{dc} = V_o = \frac{\text{Area under the curve over the full cycle}}{\text{base}}$$

where, base = full cycle time = T

$$= \frac{\int_{o}^{T} v d\theta}{T} = \frac{\int_{o}^{T} V_m \sin\theta d\theta + \int_{\pi}^{2\pi} o.d\theta}{2\pi}$$

$$= \frac{\int_{o}^{\pi} V_m \sin\theta d\theta}{2\pi} = \frac{V_m}{2\pi}\left|-\cos\theta\right|_{o}^{\pi} = \frac{V_m}{2\pi}\left|+1-(-1)\right| = \frac{V_m}{2\pi} = 0.318 V_m$$

From the above expression, we can say that the average or DC output voltage is 31.8% of the maximum AC input voltage.

Average DC Load Current

The instantaneous current i flows though the diode D and the load resistance R_L and it can be expressed as:

$$i = I_m \sin \omega t = I_m \sin\theta \quad o \le \theta \le \pi$$

$$= o\pi \le \theta \le 2\pi$$

Where $l_m = \frac{V_m}{R_f + R_L}$ as R_f is the ON state resistance of diode and R_L is load resistance.

As, $R_L \gg R_f, l_m = \frac{V_m}{R_L}$,

$$l_{av} = l_{dc} = \frac{\text{Area under curve i over the full cycle}}{\text{base}}$$

Where, base = full cycle time = T

From the above equation, it is justified that the average or DC load current is 31.8% of the maximum load current.

Single-Phase Half-Wave Uncontrolled Rectifier with RL Load and Free Wheeling Diode

In single-phase half-wave uncontrolled rectifier with RL load, the load current i is discontinuous and it is same as line current. When a freewheeling diode is connected across load, only positive half cycle supply voltage is applied across load as diode D will conduct from o to π duration and line current flows during o to π duration.

Single-phase half-wave uncontrolled rectifier with RL load and freewheeling diode is shown in Figure below. At ω t = π, conduction of diode D stops and freewheeling diode D_F starts conduction. If D is OFF and D_F scans conduction, load current flows through RL load and diode D_F, but the magnitude of load current decreases. The load current may be continuous or discontinuous depending upon β or the value of inductance L.

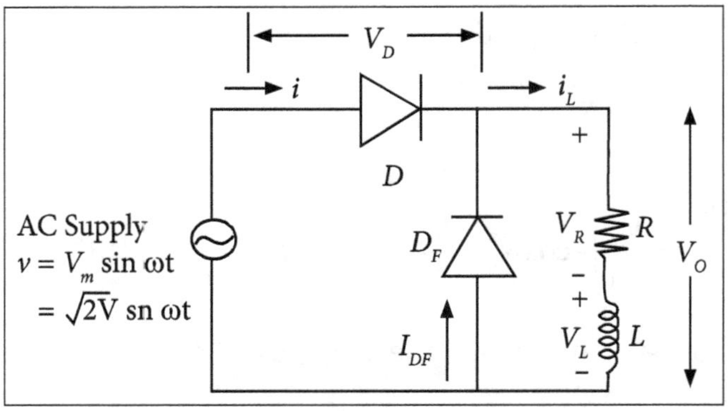

Single phase half wave controlled rectifier with RL load and freewheeling diode.

Figure below shows the voltage and current waveforms. The output voltage is same as resistive load, but the load current waveform is different. Due to the presence of freewheeling diode in the figure above, the output voltage is always positive.

The energy stored in inductance L during positive half cycle of input voltage is supplied to load resistance R through freewheeling diode. As a result the system efficiency will be improved.

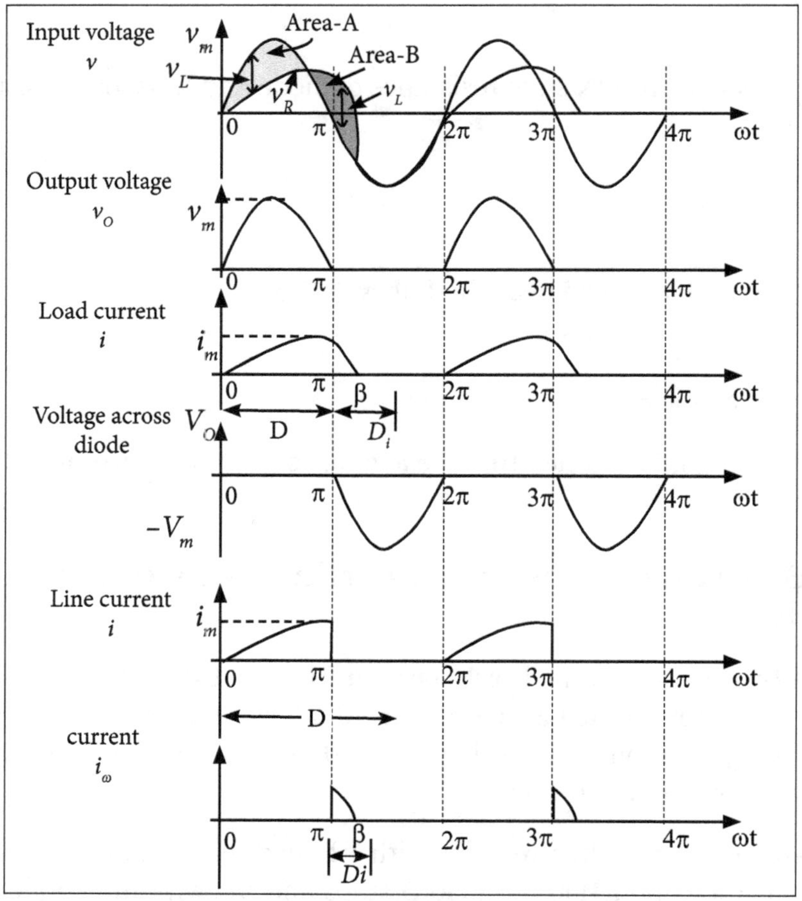

Input voltage, output voltage, load current, and voltage across diode of a single-phase half-wave rectifier circuit with RE load and freewheeling diode.

The average DC output voltage is expressed as:

$$V_o = \frac{1}{2\pi}\int_0^{\pi} \sqrt{2}V \sin \omega t d(\omega t) = \frac{\sqrt{2}V}{\pi}$$

The average load current is given by:

$$I_o = \frac{V_o}{R} = \frac{\sqrt{2}V}{\pi R}$$

3.1.1 Full Wave Controlled Rectifier with Various Kind of Loads

The single phase full wave controlled rectifier circuit combines two half wave controlled rectifiers in one single circuit so as to provide two pulse output across the load. In both, the half cycles of the input supply are utilized and converted into a uni-directional output current through the loads so as to produce a two pulse output waveform. Hence, the full wave controlled rectifier circuit is also referred to as a two pulse converter.

Types

Single phase full wave controlled rectifiers are of various types:

- The single phase full wave controlled rectifier using a center tapped transformer.
- Single phase full wave bridge controlled rectifier.
- Half controlled bridge converter (Semi converter).
- Fully controlled bridge converter (Full converter).

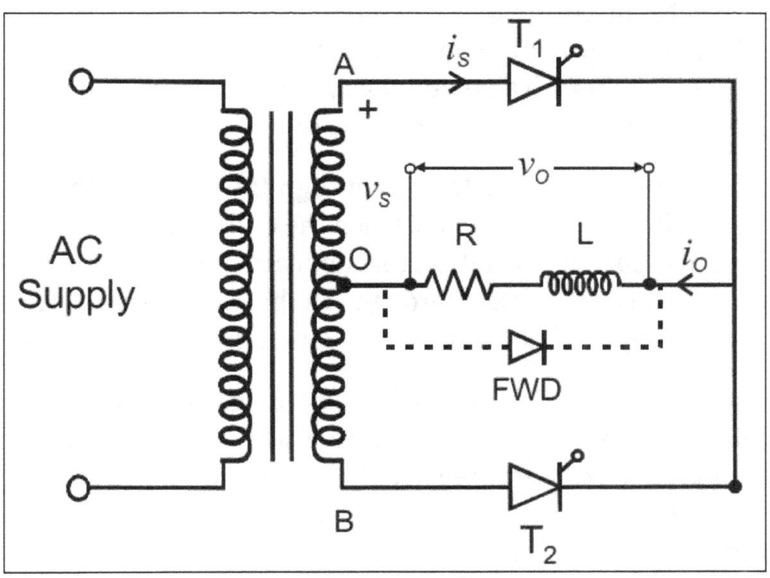

Circuit diagram of 2-pulse converters.

Here,

v_s = Supply Voltage across the upper half of the transformer secondary winding.

$v_s = v_{Ao} = V_m \sin\omega t$ = Supply voltage across the lower half of the transformer secondary winding.

This type of full wave controlled rectifier requires a center tapped transformer and two thyristors T_1 and T_2.

The input supply is fed through the main supply transformer; the primary side of the transformer is connected to the AC line voltage which is available. The secondary side of the transformer has three lines and the center point of the transformer is used as the reference point to measure the input and output voltages.

The upper half of the secondary winding and the thyristor T_1 along with the load act as a half wave controlled rectifier, the lower half of the secondary winding and the thyristor T_2 with the common load act as the second half wave controlled rectifier so as to produce a full wave load voltage waveform.

There are two types of operations possible:

- Discontinuous load current operation, which occurs for a purely resistive load or an RL load with low inductance value.
- Continuous load current operation which occurs for an RL type of load with large load inductance.

Discontinuous Load Current Operation (for Low Value of Load Inductance)

Generally, the load current is discontinuous when the load is purely resistive or when the RL load has a low value of inductance. During the positive half cycle of input supply, when the upper part of the secondary winding is at a positive potential with respect to the center point 'O', the thyristor T_1 is forward biased and it is triggered at a delay angle of α.

The load current flows through the thyristor T_1, through the load and through the upper part of the secondary winding during the period α to β, when the thyristor T_1 conducts. The output voltage across the load follows the input supply voltage that appears across the upper part of the secondary winding from $\omega t = \alpha$ to β. The load current through the thyristor T_1 decreases and drops to zero at $\omega t = \beta$, where $\beta > \pi$ or RL type of load and the thyristor T_1 naturally turns off at $\omega t = \beta$.

During the negative half cycle of the input supply, the voltage at the supply line 'A' becomes negative whereas the voltage at line 'B' becomes positive with respect to the center point 'O'. The thyristor T_2 is forward biased during the negative half cycle and it is triggered at a delay angle of $(\pi+\alpha)$. The current flows through the thyristor T_2, through the load and through the lower part of the secondary winding when T_2 conducts, during the negative half cycle the load is connected to the lower half of the secondary winding when T_2 conducts.

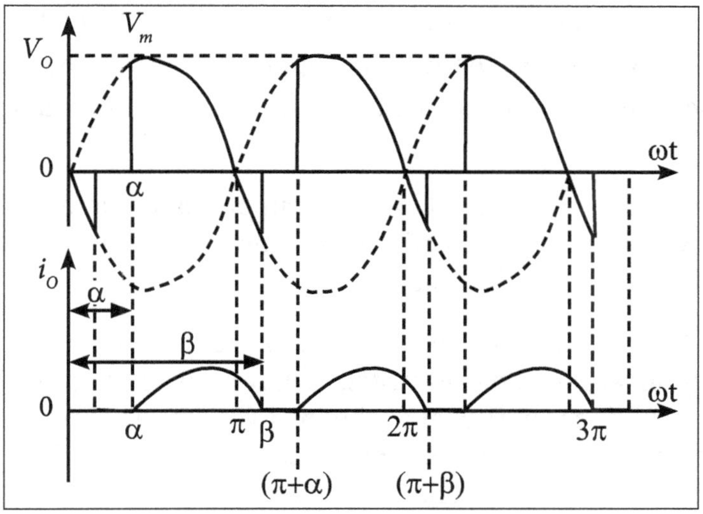

Waveform for Discontinuous Load Current Operation.

For purely resistive loads when L = 0, the extinction angle $\beta = \pi$. The load current falls to zero at $\omega t = \beta = \pi$, when the input supply voltage falls to zero at $\beta > \pi$. The load current and the load voltage waveform are in phase and there is no phase shift between the load voltage and the load current waveform in the case of a purely resistive load. The low values of load inductance, the load current would be discontinuous and the extinction angle $\beta > \pi$ but $\beta < (\pi+\alpha)$.

For large values of load inductance, the load current would be continuous and does not fall to zero. The thyristor T_1 conducts from α to $(\pi+\alpha)$, until the next thyristor T_2 is triggered. When T_2 is triggered at $\omega t = (\pi+\alpha)$, the thyristor T_1 will be reverse biased and hence T_1 turns off.

3.2 Half Controlled and Full Controlled Bridges with Passive and Active Loads

Resistive Load

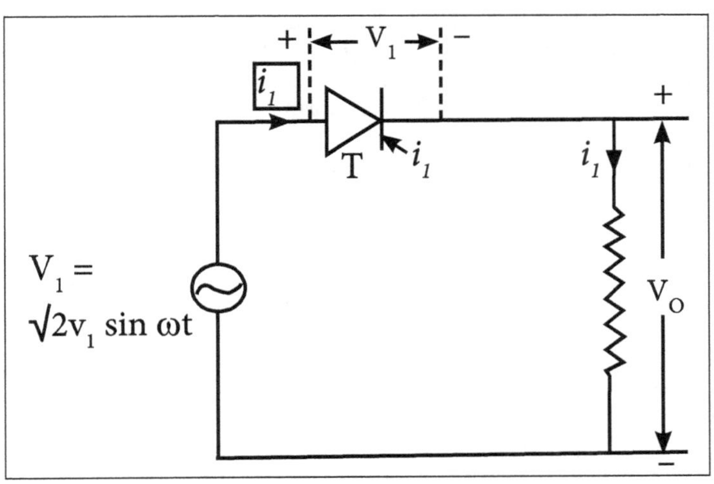

(a) Circuit diagram.

Figure (a) shows the circuit diagram of a single phase fully controlled half wave rectifier supplying a purely resistive load. At $\omega t = 0$, when the input supply voltage becomes positive, the thyristor T becomes forward biased. However, unlike a diode, it does not turn ON till gate pulse is applied at $\omega t = \alpha$.

During the period $0 < \omega t \leq \alpha$, the thyristor blocks the supply voltage and load voltage remains zero as shown in figure (b). Consequently, no load current flows during this interval. When a gate pulse is applied to the thyristor at $\omega t = \alpha$, it turns ON.

The voltage across the thyristor collapses to almost zero and full supply voltage appears across the load. From these point on-wards, the load voltage follows supply voltage. The load being purely resistive the load current i_o is proportional to the load voltage. At $\omega t = \pi$, as the supply voltage passes through the negative going zero crossing the load voltage and therefore, the load current becomes zero and tries to reverse direction.

In this process, the thyristor undergoes reverse recovery and starts blocking negative supply voltage. Therefore, the load voltage and the load current remains clamped at zero till thyristor is fired again at $\omega t = 2\pi + \alpha$. The same process repeats thereafter.

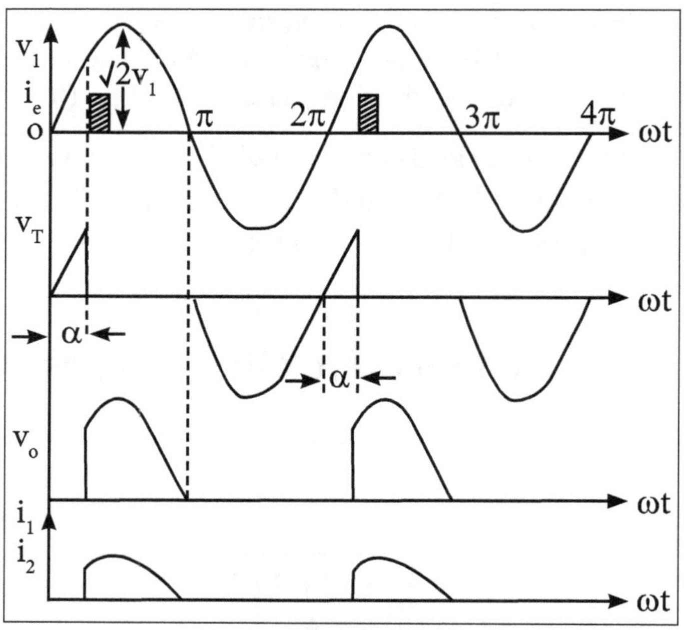

(b) Waveform

From the discussion above and Figure (b), we can write:

For $\alpha < \omega t \leq \pi$

$$v_o = v_i = \sqrt{2}\, V_i \sin\omega t \quad ...(1)$$

$$i_o = \frac{v_o}{R} = \sqrt{2}\frac{V_i}{R}\sin\omega t \quad ...(2)$$

$$v_o = i_o = 0 \text{ otherwise}$$

Therefore,

$$V_{OAV} = \frac{1}{2\pi}\int_{o}^{2\pi} v_o d\omega t = \frac{1}{2\pi}\int_{\alpha}^{\pi} \sqrt{2}V_i \sin\omega t\, d\omega t \quad ...(3)$$

$$V_{OAV} = \frac{V_i}{\sqrt{2}\pi}(1+\cos\alpha) \quad ...(4)$$

$$V_{ORMS} = \sqrt{\frac{1}{2\pi}\int_{o}^{2\pi} v_o^2 d\omega t} \quad ...(5)$$

$$= \sqrt{\frac{1}{2\pi}\int_{o}^{\pi} 2v_i^2 \sin^2 \omega t d\omega t}$$

$$= \sqrt{\frac{V_i^2}{2\pi}\int_{\alpha}^{\pi}(1-\cos 2\omega t)d\omega t}$$

$$= \sqrt{\frac{V_i^2}{2\pi}\left[\pi-\alpha+\frac{\sin 2\alpha}{2}\right]}$$

$$= \frac{V_i}{\sqrt{2}}\left(1-\frac{\alpha}{\pi}+\frac{\sin 2\alpha}{2\pi}\right)^{\frac{1}{2}}$$

$$FF_{vo} = \frac{V_{ORMS}}{V_{OAV}} = \frac{\pi\left(1-\frac{\alpha}{\pi}+\frac{\sin 2\alpha}{2\pi}\right)^{\frac{1}{2}}}{(1+\cos\alpha)} \quad ...(6)$$

Similar calculation can be done for i_o. In particulars, for pure resistive loads, $FF_{io} = FF_{vo}$.

Resistive-Inductive Load

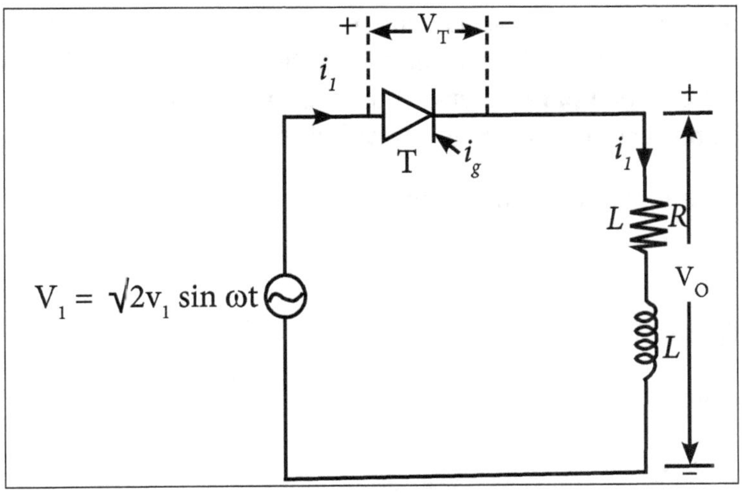

(a) Circuit diagram

Figure (a) and (b) shows the circuit diagram and waveforms of a single phase fully controlled half wave rectifier supplying a resistive inductive load. Though this circuit is hardly used in practice, its analysis does provide an useful insight into the operation of fully controlled rectifiers.

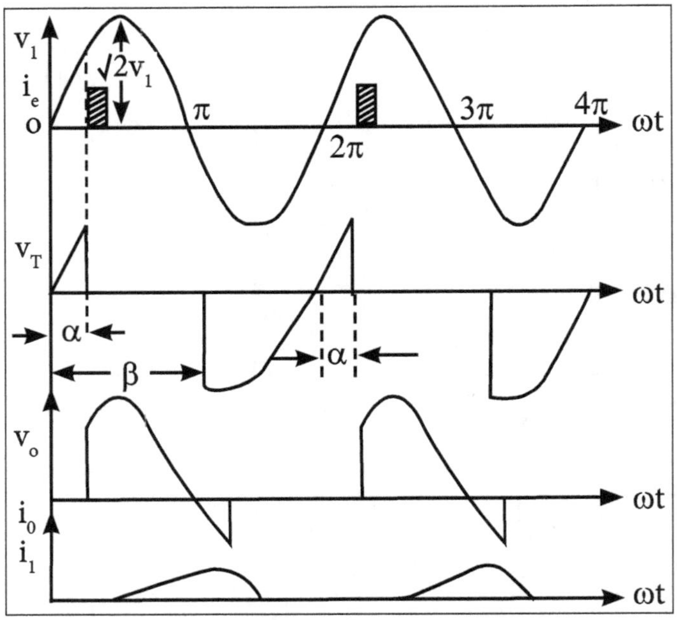

(b) Waveforms

As in the case of a resistive load, the thyristor T becomes forward biased when the supply voltage becomes positive at $\omega t = 0$. However, it does not start conduction until a gate pulse is applied at $\omega t = \alpha$. As the thyristor turns ON at $\omega t = \alpha$, input voltage appears across the load and the load current starts building up.

However, unlike a resistive load, load current does not become zero at $\omega t = \pi$, instead it continues to flow through the thyristor and the negative supply voltage appears across the load forcing the load current to decrease. Finally, at $\omega t = \beta(\beta > \pi)$ the load current becomes zero and thyristor undergoes reverse recovery.

From these point on-wards, the thyristor starts blocking the supply voltage and the load voltage remains zero until the thyristor is turned ON again in the next cycle. It is to be noted that value of β depends on the load parameters. Therefore, unlike the resistive load the average and RMS output voltage depends on the load parameters. Since the thyristors does not conduct over the entire input supply cycle this mode of the operation is called the "discontinuous conduction mode".

From the above discussion, we can writhe that:

For $\alpha \le \omega t \le \beta$

$$v_o = v_i = \sqrt{2}\,V_i \sin \omega t \qquad ...(1)$$

$v_o = 0$ otherwise

$$V_{OAV} = \frac{1}{2\pi}\int_0^{2\pi} v_o d\omega t \qquad ...(2)$$

$$=\frac{1}{2\pi}\int_{\alpha}^{\beta}\sqrt{2}\ V_i \sin\omega t\ d\omega t$$

$$=\frac{V_i}{\sqrt{2}\pi}(\cos\alpha-\cos\beta)$$

$$V_{ORMS}=\sqrt{\frac{1}{2\pi}\int_{o}^{2\pi}v_o^2 d\omega t} \qquad ...(3)$$

$$=\sqrt{\frac{1}{2\pi}\int_{\alpha}^{\beta}2v_i^2\sin^2\omega t\ d\omega t}$$

$$=\frac{V_i}{\sqrt{2}}\left(\frac{\beta-\alpha}{\pi}+\frac{\sin 2\alpha-\sin 2\beta}{2\pi}\right)^{\frac{1}{2}}$$

$$I_{OAV}=\frac{V_{OAV}}{R}=\frac{V_i}{\sqrt{2}\ \pi R}(\cos\alpha-\cos\beta) \qquad ...(4)$$

Since the average voltage drop across the inductor is zero.

However, I_{ORMS} cannot be obtained from V_{ORMS} directly. For that a closed from expression for i_o will be required. The value of β in terms of the circuit parameters could also be found from the expression of i_o.

For $\alpha \le \omega t \le \beta$

$$Ri_o + L\frac{di_o}{dt} = v_o = \sqrt{2}V_i \sin\omega t \qquad ...(5)$$

The general solution of which is given by:

$$i_o = I_o e^{-\frac{(\omega t-\alpha)}{\tan\varphi}} + \frac{\sqrt{2}V_i}{Z}\sin(\omega t\text{-}\varphi) \qquad ...(6)$$

Where $\tan\varphi=\frac{\omega L}{R}$ and $Z=\sqrt{R^2+\omega^2L^2}$

$$i_o\big|_{\omega^-=\alpha}=0$$

$$\therefore 0 = 1_o + \frac{\sqrt{2}V_i}{Z}\sin(\alpha-\varphi)$$

$$i_o=\frac{\sqrt{2}V_i}{Z}\left[\sin(\varphi-\alpha)e^{-\frac{(\omega t-\alpha)}{\tan\varphi}}+\sin(\omega t-\varphi)\right] \qquad ...(7)$$

$i_o = 0$ otherwise.

Equation (7) can be used to find out I_{ORMS}. To find out β, it is noted that

$$i_o\big|_{\omega t=\alpha} = 0$$

$$\therefore \sin(\varphi-\alpha)+e^{\frac{(\omega t-\alpha)}{\tan\varphi}} = \sin(\varphi-\beta) \qquad ...(8)$$

Equation (8) can be solved to find out β.

Single Phase Bridge Converter with RLE Load

The Figure I(a) shows the circuit diagram of a single phase fully controlled bridge converter. It is one of the most popular converter circuits and is widely used in the speed control of separately excited DC machines. Indeed, the R–L–E load shown in this figure may represent the electrical equivalent circuit of separately excited DC motor.

The single phase fully controlled bridge converter is obtained by replacing all the diode of corresponding uncontrolled converter by thyristors. Thyristors T_1 and T_2 are fired together while the T_3 and T_4 are fired 180° after T_1 and T_2. From the circuit diagram of Figure I(a), it is clear that for any load current to flow at least one thyristor from the top group (T_1, T_3) and one thyristor from the bottom group (T_2, T_4) should conduct.

It could also be argued that neither T_1T_3 nor T_2T_4 can conduct simultaneously. For example, whenever T_3 and T_4 are in the forward blocking state and a gate pulse is applied to them, they turn ON and at the same time a negative voltage is applied across T_1 and T_2 commutating them immediately. Similar argument holds for T_1 and T_2.

For the same reason, T_1T_4 or T_2T_3 cannot conduct simultaneously. Therefore, the only possible conduction modes when the current io can flow are T_1T_2 and T_3T_4. Of-coarse, it is possible that at a given moment, none of the thyristors conduct. This situation will typically occur when load current becomes zero in between the firings of T_1T_2 and T_3T_4.

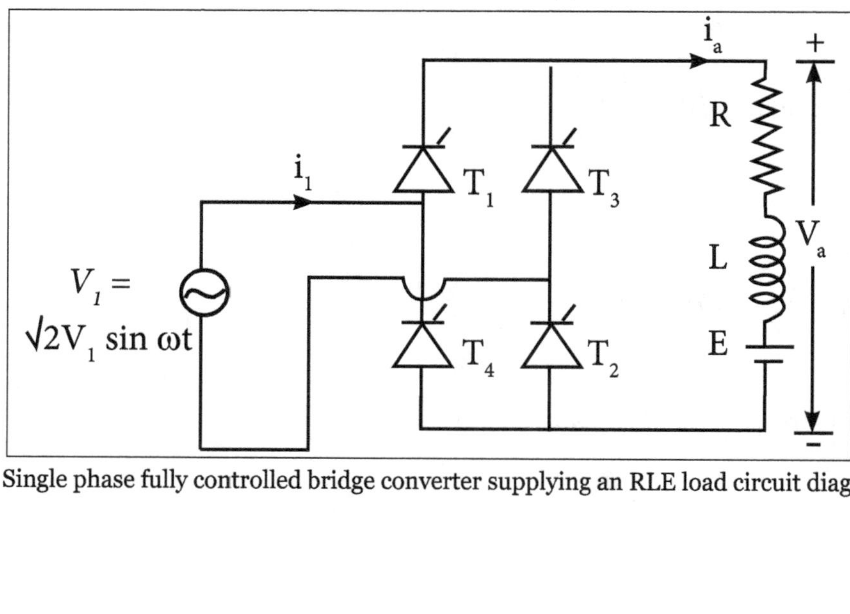

I(a) Single phase fully controlled bridge converter supplying an RLE load circuit diagram.

MODE / DEVICE	T_1	T_2	T_4	T_5	v_0
T_1T_2	0	0	$-v_1$	$-v_1$	v_1
T_3T_4	v_1	v_1	0	0	$-v_1$
NONE	$\frac{v_1 - E}{2}$	$\frac{v_1 + E}{2}$	$-\frac{v_1 + E}{2}$	$-\frac{v_1 + E}{2}$	E

I(b) Conduction table.

Once the load current becomes zero, all the thyristors remain off. In this mode, the load current remains zero. Consequently, the converter is said to be operating in discontinuous conduction mode. Figure I(b) shows the voltage across different devices and the DC output voltage during each of these conduction modes. It is to be noted that whenever T_1 and T_2 conducts, the voltage across T_3 and T_4 becomes $-v_i$. Hence, T_3 and T_4 can be fired only when v_i is negative i.e, over the negative half cycle of input supply voltage.

Similarly, T_1 and T_2 can be fired only over the positive half cycle of input supply. The voltage across the devices when none of the thyristors conduct depends on off state impedance of each device. The values listed in Figure (b) assume identical devices.

Under normal operating condition of converter, the load current may or may not remain zero over some interval of the input voltage cycle. If i_0 is always greater than zero, then converter is said to be operating in the continuous conduction mode. In this mode of operation of converter, T_1T_2 and T_3T_4 conduct for alternate half cycle of the input supply.

However, in the discontinuous conduction mode, none of the thyristors conduct over some portion of the input cycle. The load current remains zero during that period.

Operation in the Continuous Conduction Mode

i_0 never becomes zero. Hence, either T_1T_2 or T_3T_4 conducts. Figure shows the waveforms of different variables in steady state. The firing angle of the converter is α. The angle θ is given by,

$$\sin\theta = \frac{E}{\sqrt{2V_1}} \qquad ...(1)$$

It is assumed that at t = 0, T_3T_4 was conducting. As T_1T_2 are fired at ωt = α, they turn on commutating T_3T_4 immediately. T_3T_4 are again fired at ωt = π + α. Till this point, T_1T_2 conducts. The DC link voltage waveform shown below follows from this conduction diagram and the conduction table shown in Figure I(b).

It is observed that the emf source E is greater than the DC link voltage till ωt = α. Therefore, the load current i_o continues to fall till this point. Therefore, as T_1T_2 are fired at this point, v_o becomes greater than E and i_o starts increasing through R-L and E. At $\omega t = \pi - \theta$, v_o again equals E. Depending upon the load circuit parameters, i_o reaches its maximum at around this point and starts falling afterwards.

The continuous conduction mode will be possible only if i_o remains greater than zero till T_3T_4 are fired at $\omega t = \pi + \alpha$ where upon the same process repeats. The resulting i_o waveform is shown below v_o. Input AC current waveform i_i is obtained from i_o by the noting that whenever T_1T_2 conducts $i_i = i_o$ and $i_i = - i_o$ whenever T_3T_4 conducts.

The last waveform shows the typical voltage waveform across the thyristor T_1. It is to be noted that when thyristor turns off at $\omega t = \pi + \alpha$, the negative voltage is applied across it for a duration of $\pi - \alpha$. The thyristor must turn off during this interval for the successful operation of the converter.

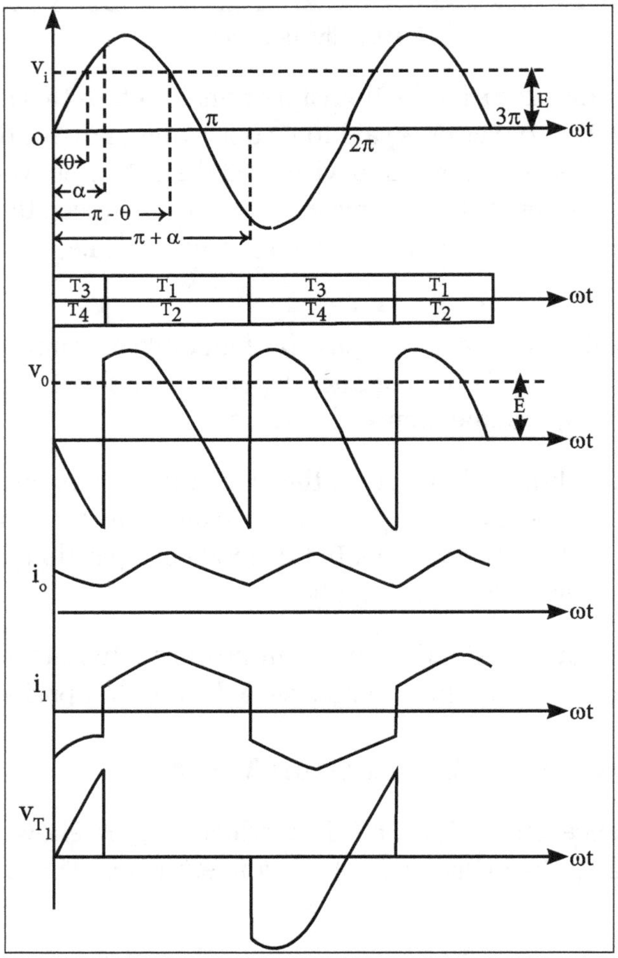

II. Waveforms of Continuous conduction mode.

It is noted that the DC voltage waveform is periodic over half the input cycle. Therefore, it can be expressed in a Fourier series as follows.

$$v_o = V_{OAV} + \sum_{n=1}^{a}[v_{an}\cos 2n\omega t + v_{bn}\sin 2n\omega t] \qquad ...(2)$$

Where,

$$V_{OAV}=\frac{1}{\pi}\int_{a}^{\pi+a}v_{o}dwt=\frac{2\sqrt{2}}{\pi}V_{i}\cos\alpha \qquad ...(3)$$

$$v_{an}=\frac{2}{\pi}\int_{o}^{\pi}v_{o}\cos 2n\omega t\, d\omega t=\frac{2\sqrt{2}}{\pi}V_{i}\left[\frac{\cos(2n+1)\alpha}{2n+1}-\frac{\cos(2n-1)\alpha}{2n-1}\right] \qquad ...(4)$$

$$v_{bn}=\frac{2}{\pi}\int_{o}^{\pi}v_{o}\sin 2n\omega t\, d\omega t=\frac{2\sqrt{2}}{\pi}V_{i}\left[\frac{\sin(2n+1)\alpha}{2n+1}-\frac{\sin(2n-1)\alpha}{2n-1}\right] \qquad ...(5)$$

Therefore, the RMS value of the nth harmonic:

$$V_{OnRMS}=\frac{1}{\sqrt{2}}\sqrt{v_{an}^{2}+v_{bn}^{2}} \qquad ...(6)$$

RMS value of v_o can be completed directly from:

$$V_{OnRMS}=\sqrt{\frac{1}{\pi}\int_{\alpha}^{\pi+\alpha}v_{o}^{2}d\omega t}=V_{i} \qquad ...(7)$$

Fourier series expression of v_o is important because it provides a simple method of estimating individual and total RMS harmonic current injected into the load as follows:

The impedance offered by the load at nth harmonic frequency is given by:

$$Z_{n}=\sqrt{R^{2}+(2nwL)^{2}} \qquad ...(8)$$

$$I_{onRMS}=\frac{V_{onRMS}}{Z_{n}};I_{OHRMS}=\left[\sum_{n=1}^{\alpha}I_{onRMS}^{2}\right]^{\frac{1}{2}} \qquad ...(9)$$

From (4) – (9), it can be argued that in an inductive circuit $I_{onRMS}\rightarrow 0$ as fast as $1/n^2$. So in practice, it will be sufficient to consider only first few harmonics to obtain a reasonably accurate estimate of the I_{OHRMS} from equation (9). This method will be useful, for example, while calculating the required current derating of a DC motor to be used with such a converter.

Therefore, to obtain the current rating of the device to be used, it is necessary to find out a closed form expression of i_o. This will also help to establish the condition under which converter will operate in the continuous conduction mode.

We observe that voltage waveform and hence the current waveform is periodic over an interval π. Therefore, finding out an expression for i_o over any interval of length π will be sufficient. We choose the interval $\alpha \le \omega t \le \pi + \alpha$. In this interval:

$$L\frac{di_{o}}{dt}+Ri_{o}+E=\sqrt{2}V_{i}\sin\omega t \qquad ...(10)$$

The general solution is given by:

$$i_o = Ie^{-\frac{\omega t-\alpha}{\tan\varphi}} + \frac{\sqrt{2}V_i}{Z}\left[\sin(\omega t-\varphi) - \frac{\sin\theta}{\cos\varphi}\right] \quad ...(11)$$

Where,

$$Z = \sqrt{R^2+\omega^2L^2}; \tan\varphi = \frac{\omega L}{R}; E = \sqrt{2}V_i \sin\theta; R = Z\cos\varphi$$

Now at steady state $i_o|_{\omega t=a} = i_o|_{\omega t=\pi+a}$, since i_o is periodic over the chosen interval. Using this boundary condition we obtain:

$$i_o = \frac{\sqrt{2}V_i}{Z}\left[\frac{2\sin(\varphi-\alpha)}{1-e^{-\frac{\pi}{\tan\varphi}}}e^{-\frac{\omega t-\alpha}{\tan\varphi}} + \sin(\omega t-\varphi) - \frac{\sin\theta}{\cos\varphi}\right] \quad ...(12)$$

The input current i_i is related to i_o as follows:

$$i_i = i_o \text{ for } \alpha\ \omega \le\le t\ \pi+\alpha \quad ...(13)$$

$$i_i = -i_o \text{ otherwise.}$$

Figure (III) shows the waveform of i_i in relation to the v_i waveform. However, using actual expression for i_i will lead to exceedingly complex calculation. Significant simplification could be made by replacing i_o with its average value I_o. This will be justified provided that load is highly inductive and the ripple on i_o is negligible compared to I_o.

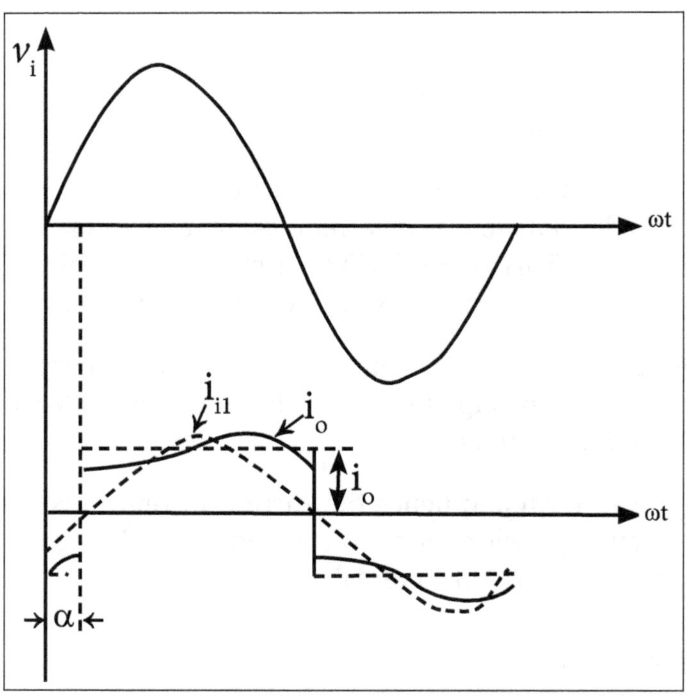

III. The input current and its fundamental component.

Under this assumption, the idealized waveform of i_i becomes a square wave with transitions at $\omega t = \alpha$ and $\omega t = \alpha + \pi$ as shown in Figure III. i_{i1} is fundamental component of this idealized i_i Evidently input current displacement factor defined as the cosine of the angle between input voltage (v_i) and the fundamental component of input current (i_{i1}) waveforms is cosα (lagging).

It can be shown that:

$$I_{iIRMS} = \frac{2\sqrt{2}}{\pi} I_o \quad ...(14)$$

$$I_{iRMS} = I_o \quad ...(15)$$

Therefore the input current distortion factor $= \frac{I_{iIRMS}}{I_{iRMS}} = \frac{2\sqrt{2}}{\pi}$ ...(16)

The input power factor $= \frac{\text{Actual Power}}{\text{Apparent Power}} = \frac{V_i I_{iIRMS} \cos\alpha}{V_i I_{iRMS}}$

$$= \frac{2\sqrt{2}}{\pi} \cos\alpha \text{ (lagging)} \quad ...(17)$$

Therefore, the rectifier appears as a lagging power factor load to the input AC system.

Larger the 'α', poorer is the power factor. The input current i_i also contain significant amount of harmonic current (3rd, 5th, etc) and therefore appears as a harmonic source to the utility. Exact composition of the harmonic currents can be obtained by Fourier series analysis of i_i.

Operation in the Discontinuous Conduction Mode

In figure III, the voltage across the R and L component of the load is negative in the region $\pi - \theta \leq \omega t \leq \pi + \alpha$. Therefore, i_o continues to decrease till a new pair of thyristor is fired at $\omega t = \pi + \alpha$. Now if the value of R, L and E are such that the i_o becomes zero before $\omega t = \pi + \alpha$, the conduction becomes discontinuous.

Obviously, then at the boundary between continuous and discontinuous conduction the minimum value of i_o which occurs at $\omega t = \alpha$ and $\omega t = \pi + \alpha$ will be zero. Putting this condition in (12), we obtain the condition for continuous conduction as:

$$\frac{2\sin(\varphi - \alpha)}{1 - e^{-\frac{\pi}{\tan\varphi}}} - \sin(\varphi - \alpha) - \frac{\sin\theta}{\cos\varphi} \geq 0 \quad ...(18)$$

Figure IV shows waveforms of different variables on the boundary between continuous and discontinuous conduction modes and in the discontinuous conduction mode. It should be stressed that on the boundary between continuous and discontinuous conduction modes, the load current is still continuous. Hence, all the analysis of continuous conduction mode applies to this case as well.

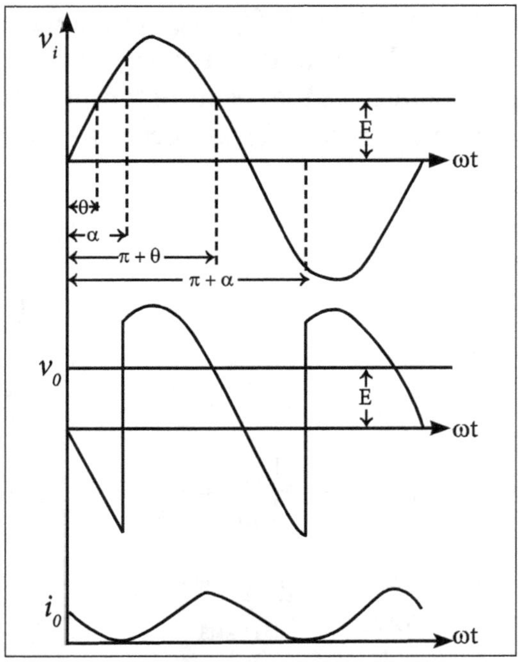

IV(a) On the boundary between continuous and discontinuous conduction modes.

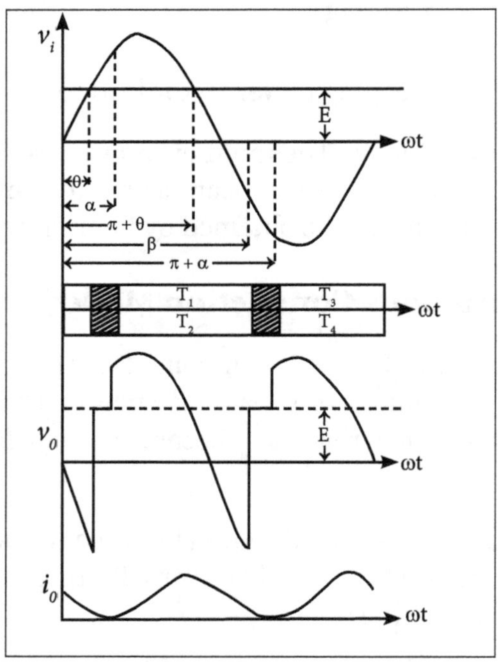

IV(b) The discontinuous conduction mode.

However in the discontinuous conduction mode i_0 remains zero for certain interval. During this interval none of the thyristors conduct. These intervals are shown by the hatched lines in the conduction diagram of Figure IV (b). In this conduction mode, i_0 starts rising from zero as T_1T_2 are fired at $\omega t = \alpha$.

The load current continues to increase till $\omega t = \pi - \theta$. After this, output voltage v_0 falls below the emf E and i_0 decreases till $\omega t = \beta$ when it becomes zero. Since the thyristors cannot conduct current in the reverse direction i_0 remains at zero till $\omega t = \pi + \alpha$ when T_3 and T_4 are fired.

During the interval $\beta \leq \omega t \leq \pi + \alpha$, none of the thyristors conduct. During this period v_o attains the value E. Performance of the rectifier such as V_{OAV}, V_{ORMS}, I_{OAV}, I_{ORMS} etc., can be found in terms of α, β and θ. For example,

$$V_{OAV} = \frac{1}{\pi}\int_{\alpha}^{\pi+\alpha} v_o \, d\omega t = \frac{1}{\pi}\left[\int_{\alpha}^{\beta} \sqrt{2}V_i \sin\omega t \, d\omega t + \int_{\beta}^{\pi+\alpha} \sqrt{2}V_i \sin\theta \, d\omega t\right] \quad ...(19)$$

$$V_{OAV} = \frac{\sqrt{2}V_i}{\pi}\left[\cos\alpha - \cos\beta + \sin\theta(\pi + \alpha - \beta)\right] \quad ...(20)$$

$$I_{OAV} = \frac{V_{OAV} - E}{R} = \frac{V_{OAV} - \sqrt{2}V_i \sin\theta}{Z\cos\varphi} \quad ...(21)$$

$$I_{OAV} = \frac{\sqrt{2}V_i}{\pi Z\cos\varphi}\left[\cos\alpha - \cos\beta + \sin\theta(\alpha - \beta)\right] \quad ...(22)$$

It is observed that the performance of the converter is strongly affected by the value of β. The value of β in terms of the load parameters (i.e, θ, φ and Z) and α can be found as follows.

In the intervel $\alpha \leq \omega t \leq \beta$

$$L\frac{di_o}{dt} + Ri_o + E = \sqrt{2}V_i \sin\omega t \quad ...(23)$$

$$i_o\big|_{\omega t=\alpha} = 0$$

From which the solution of i_o can be written as:

$$i_o = \frac{\sqrt{2}V_i}{Z}\left[\sin(\varphi - \alpha)e^{-\frac{\omega t - \alpha}{\tan\varphi}} - \frac{\sin\theta}{\cos\varphi}\left\{1 - e^{-\frac{\omega t - \alpha}{\tan\varphi}}\right\} + \sin(\omega t - \varphi)\right] \quad ...(24)$$

$$i_o\big|_{\omega t=\beta} = 0$$

$$\text{Sin}(\varphi - \alpha)e^{\frac{\alpha - \beta}{\tan\varphi}} - \frac{\sin\theta}{\cos\varphi}\left[1 - e^{-\frac{\omega t - \alpha}{\tan\varphi}}\right] + \sin(\beta - \varphi) = 0 \quad ...(25)$$

Given φ, α and θ, the value of β can be found by solving equation 25.

3.2.1 Input Line Current Harmonics and Power Factor

Harmonic Analysis for Input Current Waveform in a System with a Large Load Inductance

Figure I show a simplified schematic of the controlled rectifier circuit. In the figure, four thyristors are connected in a full bridge configuration with an AC input and a resistive load output. The firing angle (α) of thyristors could be controlled through their gates to chop the input current on the positive and negative cycles.

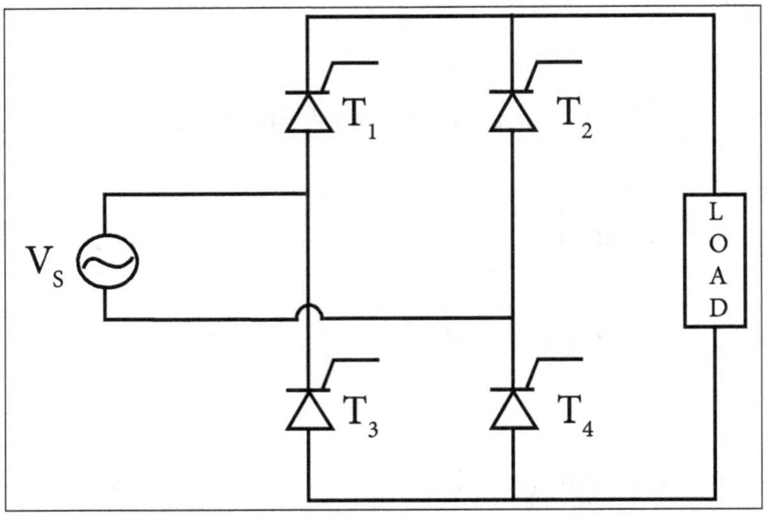

I. A simplified schematic of the Thyristor controlled circuit.

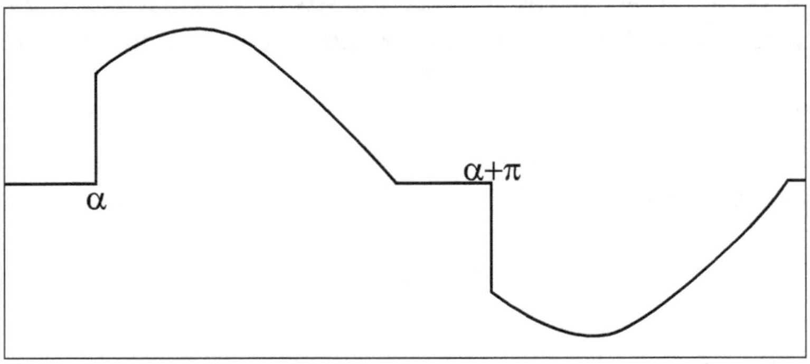

II. Input current waveform used for Fourier analysis.

Using Figure II, which shows input current waveform, the Fourier series component can be expressed as:

$$a_n = \frac{1}{\pi}\int_0^{2\pi} f(\omega t)\cos(n\omega t)d(\omega t)(1)$$

$$b_n = \frac{1}{\pi}\int_0^{2\pi} f(\omega t)\sin(n\omega t)d(\omega t)$$

Since in obtaining the THD, this analysis is only concerned with the fundamental component (n = 1), the above equations should only be used for n equal to one. The derivation of a_1 and b_1 is summarized below:

$$\pi a_1 = \int_0^{\pi} \sin(\omega t)\cos(\omega t)d(\omega t)$$

$$+\int_{\pi+\alpha}^{2\pi} \sin(\omega t)\cos(\omega t)d(\omega t)$$

$$= \frac{\sin^2(\omega t)}{2}\Bigg|_{\alpha}^{\pi} + \frac{\sin^2(\omega t)}{2}\Bigg|_{\pi+\alpha}^{2\pi}$$

$$2\pi a_1 = -2\sin^2(\alpha)$$

$$\boxed{a_1 = -\frac{1}{\pi}\sin^2(\alpha)}$$

$$\pi b_1 = \int_{\alpha}^{\pi}\sin(\omega t)\sin(\omega t)d(\omega t)$$

$$+\int_{\pi+\alpha}^{2\pi}\sin(\omega t)\sin(\omega t)d(\omega t)$$

$$2\pi b_1 = \int_{\alpha}^{\pi}(1-\cos(2\omega t))d(\omega t)$$

$$+\int_{\pi+\alpha}^{2\pi}(1-\cos(2\omega t))d(\omega t)$$

$$= t - \frac{\sin(2\omega t)}{2}\Bigg|_{\alpha}^{\pi} + t - \frac{\sin(2\omega t)}{2}\Bigg|_{\pi+\alpha}^{2\pi}$$

$$= \pi - \alpha - \frac{\sin(2\pi)}{2} + \frac{\sin(2\alpha)}{2} + 2\pi - \pi - \alpha$$

$$-\frac{\sin(4\pi)}{2} + \frac{\sin(2\pi+2\alpha)}{2}$$

$$\boxed{b_1 = 1 - \frac{\alpha}{\pi} + \frac{\sin(2\alpha)}{2\pi}}$$

Since both a_1 and b_1 are magnitudes and at the same frequency, they must be changes to rms values and their geometric mean must be calculated.

$$f_{1(rms)} = \sqrt{\frac{a_1^2 + b_1^2}{2}}$$

The total rms value is calculated from:

$$f_{tot(rms)} = \sqrt{1 - \frac{\alpha}{\pi} + \frac{\sin(2\alpha)}{2\pi}}\,\frac{1}{\sqrt{2}}$$

while the THD can be shown as:

$$THD = \frac{\sqrt{f_{tot(rms)}^{2} - f_{1(rms)}^{2}}}{f_{1(rms)}}$$

The theoretical THD at different firing angles (α) is shown in the below table.

Table: Summary of total harmonic distortion analysis.

Input Current THD (%)			
Firing angle	Fourier analysis	simulated	Experimental
5.14	1.21	1.54	8.24
10	3.21	3.34	9.81
15	5.75	5.52	11.24
30	15.11	14.76	15.59
45	25.93	25.6	25.27
60	37.74	37.59	34.03
75	50.63	50.6	49.03
90	65.05	65.18	62.93
105	81.86	82.18	64.74
120	102.63	103.24	116.89
135	130.58	131.73	143.1
150	173.92	168.27	179.05

3.2.2 Inverter Mode of Operation

Inverter

The device that converts DC power into AC power at desired output voltage and frequency is called an inverter.

Single Phase Voltage Source Inverters

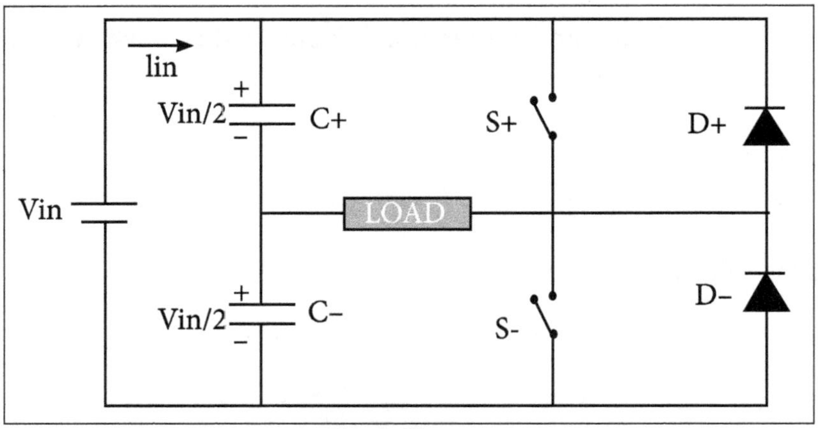

Circuit diagram.

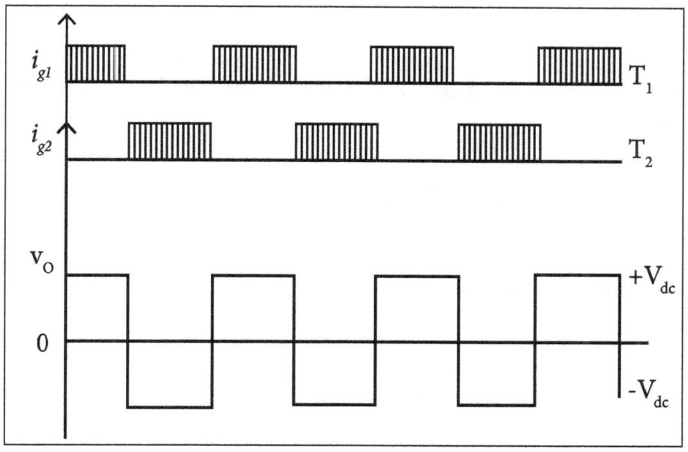

Waveform.

$$V_o(rms) = \frac{1}{T_o/2}\int_0^{T_o/2} \frac{V_S^2}{4} dt = \frac{V_S}{2}$$

$$V_{O=\frac{a_O}{2}} + \sum_{n=1}^{\infty}\left(a_n \cos(n\omega t) + b_n \sin(n\omega t)\right)$$

Due to symmetry along x-axis:

$$a_o = 0,\ a_n = 0$$

$$b_n = \frac{4V_S}{n\pi}$$

The instantaneous output voltage is:

$$\tilde{o}_o = \sum_{n=1,3,5\ldots}^{\infty} \frac{2V_S}{n\pi}\sin(nwt)$$

$$= 0,\ n = 2,4\ldots.$$

The rms value of the fundamental output voltage is:

$$V_{o1} = \frac{2V_S}{\sqrt{2\pi}} = 0.45\ V_S$$

So if, $V_O \sum_{n=1,3,5\ldots}^{\infty} \frac{2V_S}{n\pi}\sin(nwt)$

$$= \sum_{n=1,3,5\ldots}^{\infty} \frac{2V_S}{n\pi\sqrt{R^2 + (n\omega L)^2}} \sin(n\omega t - \theta_n)$$

$$P_{o1} = (I_{o1})^2 R = \left[\frac{2V_S}{\sqrt{2\pi}\sqrt{R^2 + (\omega L)^2}}\right]^2 R$$

DC Supply Current

Assuming a loss less inverter, the AC power absorbed by the load must be equal to the average power supplied by the DC source.

$$\int_0^T i_s(t)dt = \frac{1}{V_S}\int_0^T \sqrt{2}V_{01}\sin(\omega t)\sqrt{2}I_o\sin(\omega t-\theta_1)dt = I_S$$

V_{01} = Fundamental rms output output voltage.

I_o = RMS load current.

θ_1 = The load angle at the fundamental frequency.

Single Phase Full Bridge Inverter

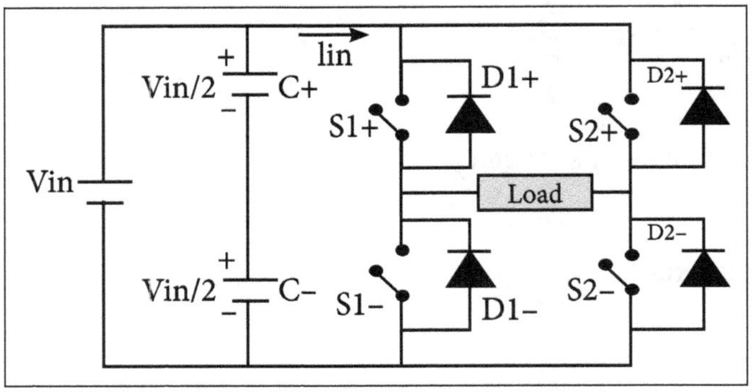

Single Phase Full Bridge Inverter.

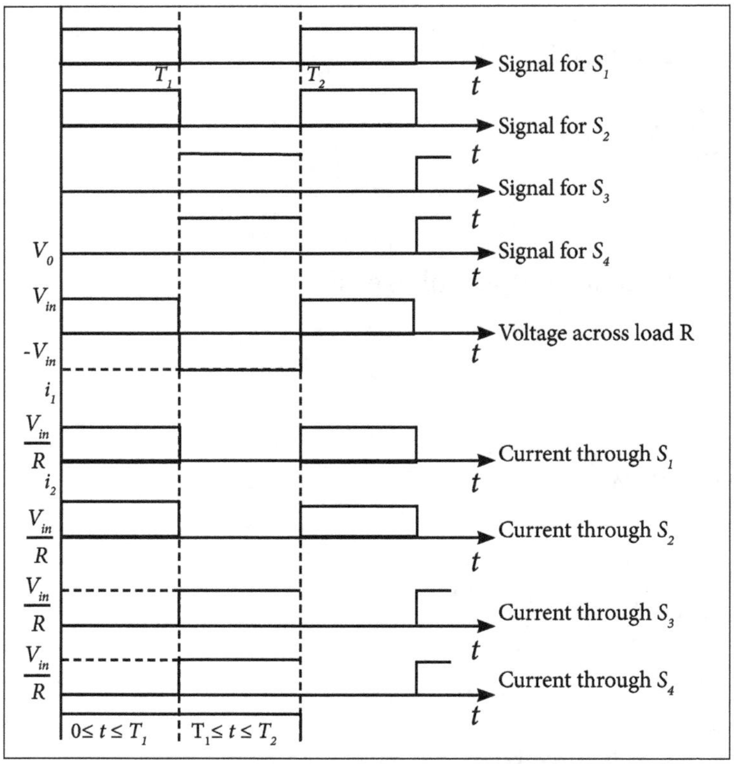

Waveform for Single Phase Full Bridge Inverter.

For $n = 1$, $V_1 = \frac{4V_S}{\sqrt{2}V_S} = 0.9\ V_S$ (The rms of fundamental)

Instantaneous load current io for an RL load:

$$i_o = \sum_{n=1,3,5...}^{\infty} \frac{4V_S}{n\pi\sqrt{2R^2 + (n\omega L)^2}} \sin(n\omega t - \theta_n)$$

$$\theta_n = \tan^{-1}\left(\frac{n\omega L}{R}\right)$$

The rms output voltage is:

$$V_o = \left(\frac{2}{T_o}\int_o^{T/2} V_S^2\right)^{1/2} = V_S$$

The instantaneous output voltage in a Fourier series is:

$$\tilde{o}_o = \sum_{n=1,3,5...}^{\infty} \frac{4V_S}{n\pi} \sin(n\omega t)$$

Three Phase Inverter

When three single-phase inverters are connected in parallel, a three phase inverter is formed. The gating signal has to be displaced by 120°with respect to each other so as achieve three phase balanced voltages.

A 3-phase output can be achieved from a configuration of six transistors and six diodes. Two types of control signal can be applied to transistors; they are such as 180° or 120° conduction.

180-degree Conduction

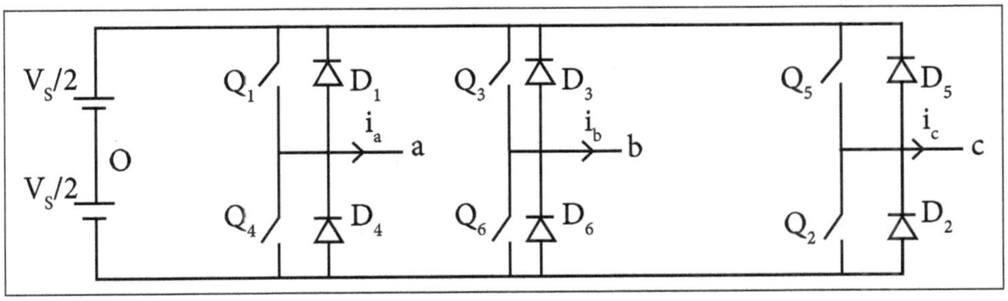

Three Phase Inverter Circuit.

When Q_1 is switched ON, terminal a is connected to the positive terminal of DC input voltage.

When Q_4 is switched ON, terminal a is brought to negative terminal of the DC source.

There are 6 modes of operation in a cycle and the duration of each mode is 60°. The conduction sequence of transistors is 123,234,345,456,561,612. The gating signals are shifted from each other by 60° to get 3-ϕ balanced voltages.

Switching States for Three Phase Voltage Inverters

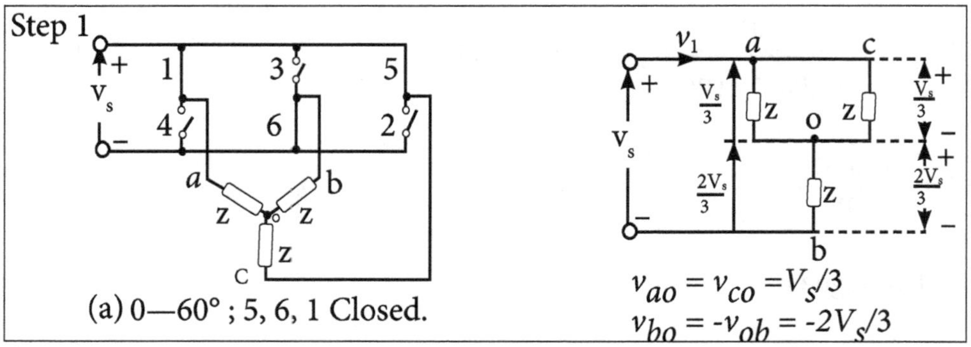

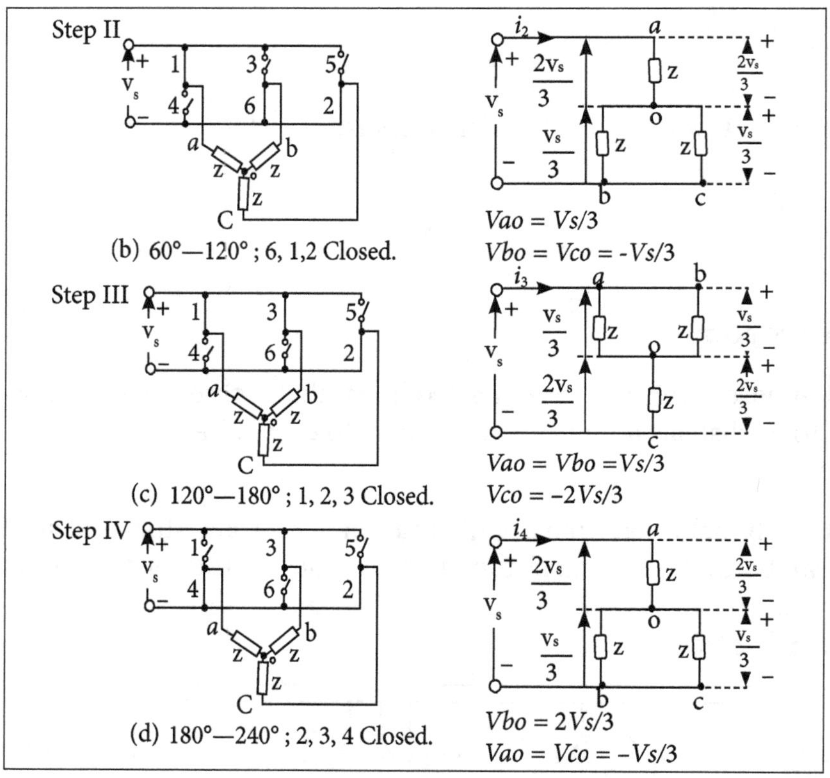

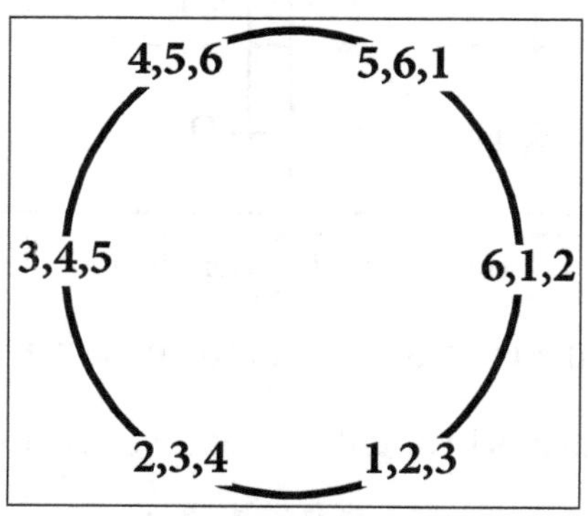

Switching states for the three phase voltage inverters - 180 degree mode.

V_{RN}	V_{YN}	V_{BN}	V_{RY}	V_{YB}	V_{BR}	V_1
$\frac{V}{3}$	$\frac{-2V}{3}$	$\frac{V}{3}$	V_{ac}	$-V_{dc}$	0	$\frac{2}{\sqrt{3}}(330^\circ)$
$\frac{2V}{3}$	$\frac{-V}{3}$	$\frac{-V}{3}$	V_{dc}	0	$-V_{dc}$	$\frac{2}{\sqrt{3}}(30^\circ)$
$\frac{V}{3}$	$\frac{V}{3}$	$\frac{-2V}{3}$	0	V	−V	$\frac{2}{\sqrt{3}}(90^\circ)$
$\frac{-V}{3}$	$\frac{2V}{3}$	$\frac{-V}{3}$	−V	V	0	$\frac{2}{\sqrt{3}}(150^\circ)$
$\frac{-2V}{3}$	$\frac{V}{3}$	$\frac{V}{3}$	−V	0	0	$\frac{2}{\sqrt{3}}(210^\circ)$
$\frac{-V}{3}$	$\frac{-V}{3}$	$\frac{2V}{3}$	0	−V	0	$\frac{2}{\sqrt{3}}(270^\circ)$

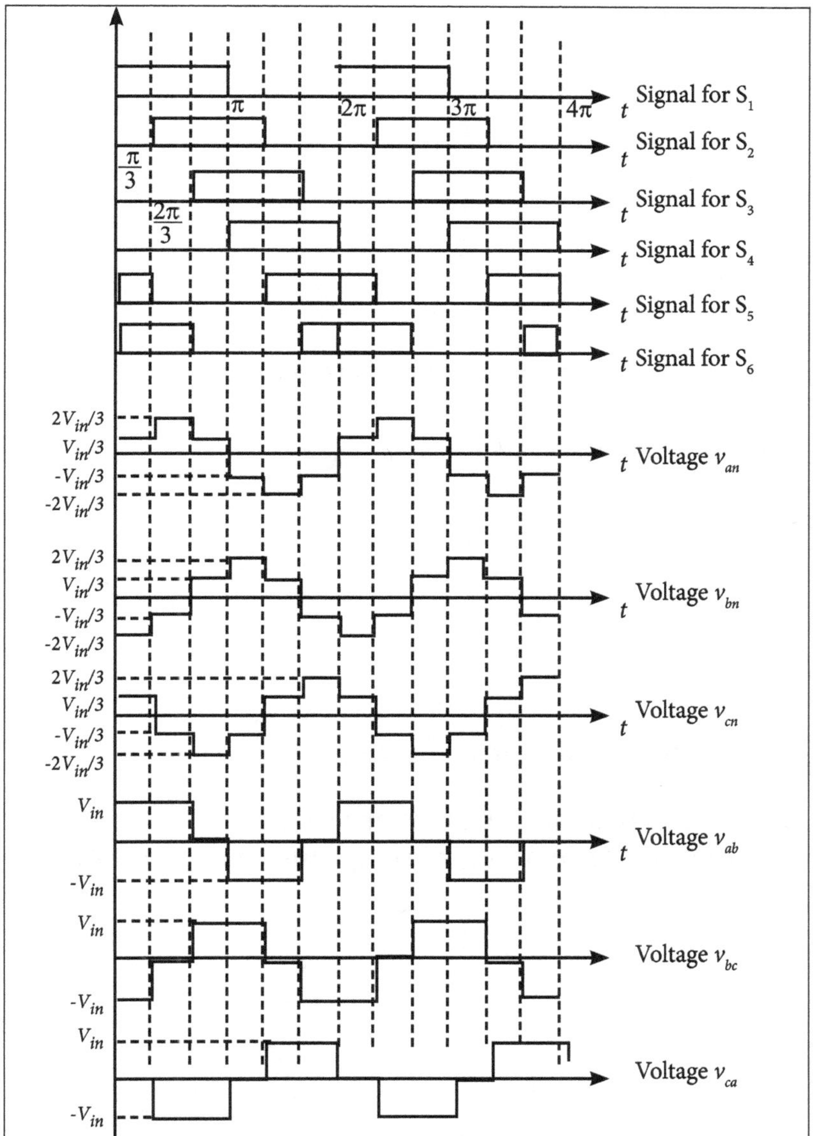

Voltage waveform rectifiers load for 180°

Fourier Analysis

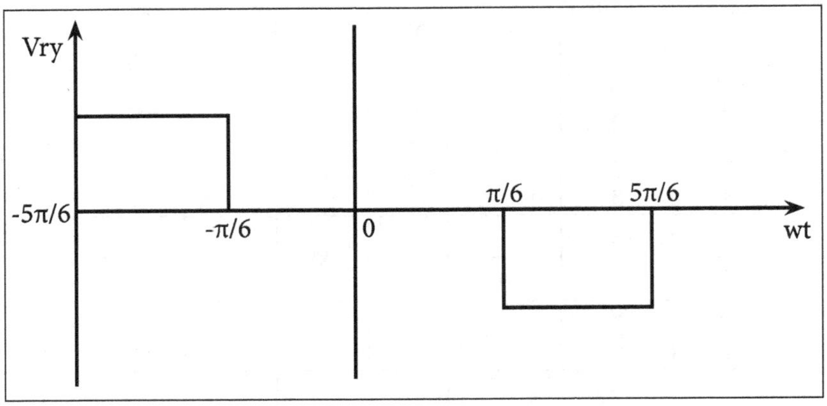

If we go for harmonic analysis:

$$V_{RY} = \sum_{n=1,3,5\ldots}^{\infty} \frac{4V}{n\pi} \sin\frac{n\pi}{3} \sin n\left(\omega t + \pi/6\right)$$

$$V_{YB} = \sum_{n=1,3,5\ldots}^{\infty} \frac{4V_S}{n\pi} \sin\frac{n\pi}{3} \sin n\left(\omega t - \pi/2\right)$$

$$V_{BR} = \sum_{n=1,3,5\ldots}^{\infty} \frac{4V_S}{n\pi} \sin\frac{n\pi}{3} \sin n\left(\omega t - \pi/6\right)$$

All even harmonics are zero & all triple n harmonics are zero.

The rms nth component of the line voltage is:

$$VBR = \frac{4V}{\sqrt{2}n\pi} \sin\frac{n\pi}{3} = \frac{4V}{\sqrt{2}\pi} \sin(60)$$

For $n = 1$

$$= 0.7797\, V_S$$

Three Phase 120° Mode Voltage Source Inverter

The circuit diagram is same as that for 180° mode of conduction. Here each thyristor conducts for 120°. There are 6 steps each of 60° duration, for completing one cycle of AC output voltage.

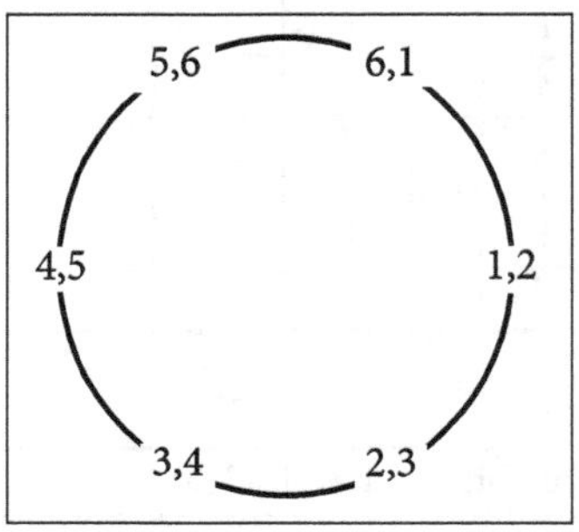

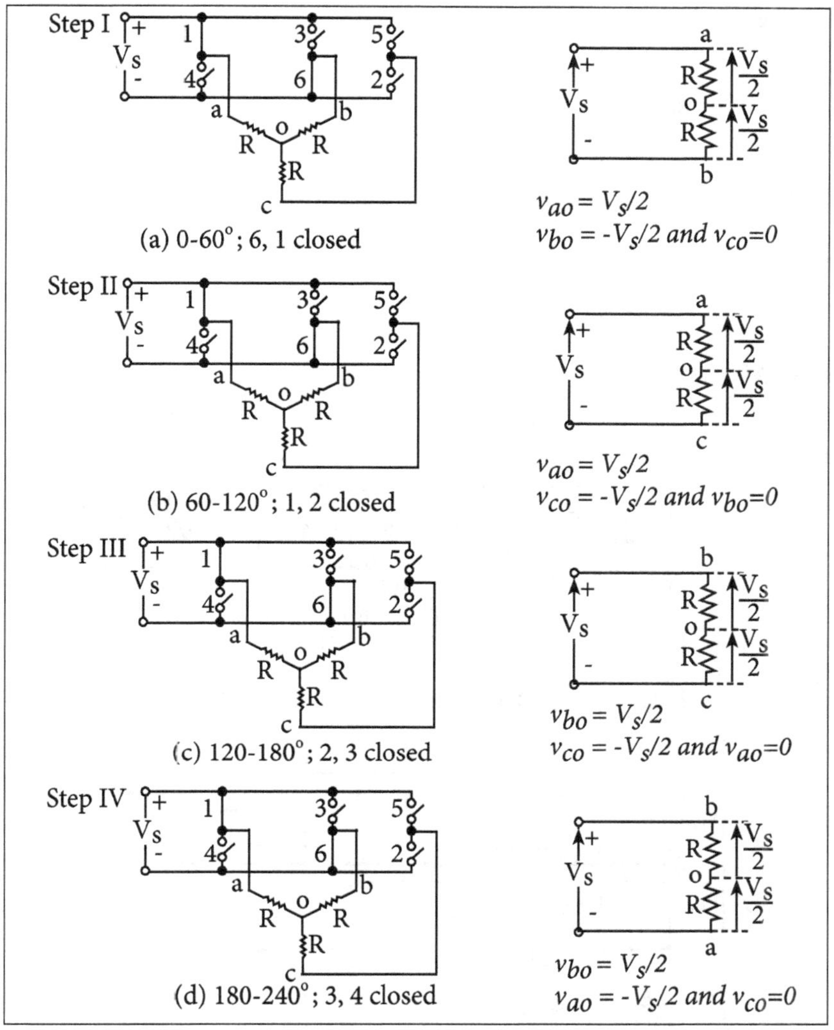

Three phase 120° mode VSI.

Step: 1 6, 1 conducting:

$$V_{an} = \frac{V_S}{2},\ V_{yn} = \frac{-V_S}{2},\ V_{cn} = 0$$

Step: 2 1, 2 conducting:

$$V_{an} = \frac{V_S}{2},\ V_{bn} = 0,\ V_{cn} = \frac{-V_S}{2}$$

Step: 3 2, 3 conducting:

$$V_{an} = 0,\ V_{bn} = \frac{V_S}{2},\ V_{cn} = \frac{-V_S}{2}$$

Step: 4 3, 4 conducting:

$$V_{an} = \frac{-V_S}{2},\ V_{bn} = \frac{V_S}{2},\ V_{cn} = 0$$

Step: 5 4, 5 conducting:

$$V_{an} = \frac{-V_S}{2}, V_{yn} = 0, V_{bn} = \frac{V_S}{2}$$

Step: 6 5, 6 conducting:

$$V_{an} = 0, V_{bn} = \frac{-V_S}{2}, V_{cn} = \frac{V_S}{2}$$

120° Conduction Mode

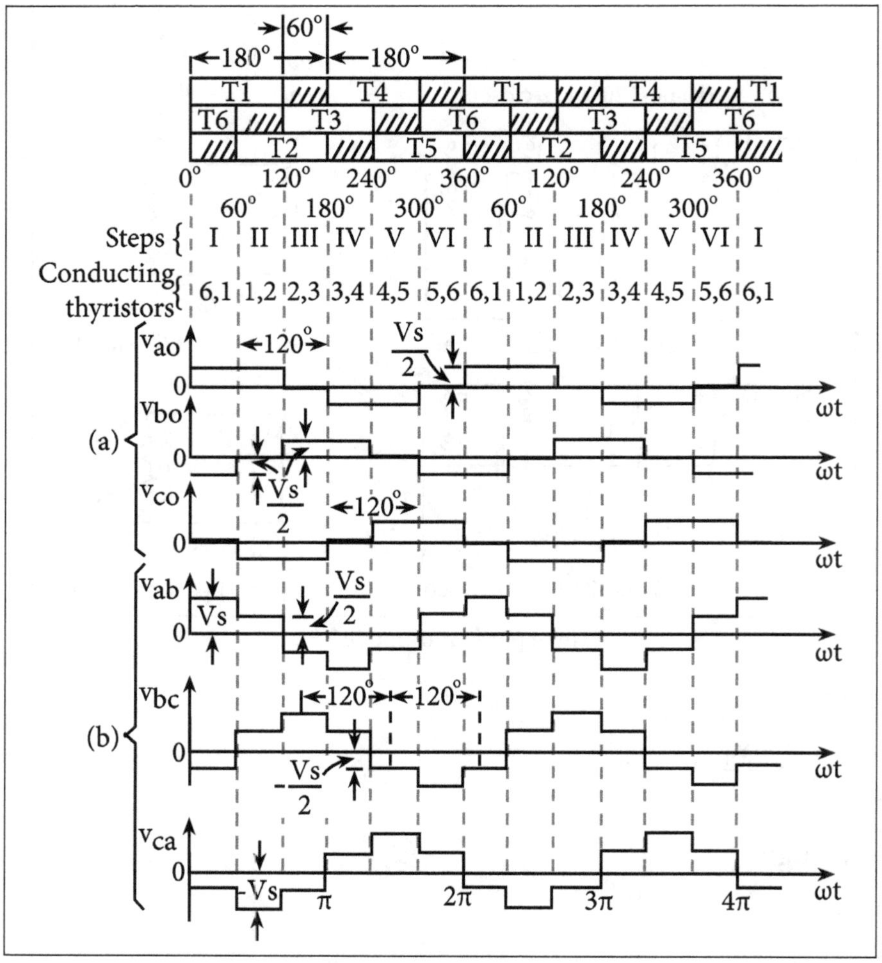

Waveform of three phase voltage source inverter at 120° conduction mode.

Step	Thyristor conduction	V_{Rn}	V_{Yn}	V_{Bn}	$\vec{V}$
1	6,1	$V_{s/2}$	$-V_{s/2}$	0	$\frac{\sqrt{3}V_S}{2}(-30^\circ)$
2	1,2	$V_{s/2}$	0	$-V_{s/2}$	$\frac{\sqrt{3}V_S}{2}(30^\circ)$

3	2,3	0	$V_s/2$	$-V_s/2$	$\frac{\sqrt{3}V_s}{2}(90^\circ)$
4	3,4	$-V_s/2$	$V_s/2$	0	$\frac{\sqrt{3}V_s}{2}(150^\circ)$
5	4,5	$-V_s/2$	0	$V_s/2$	$\frac{\sqrt{3}V}{2}(210^\circ)$
6	5,6	0	$-V_s/2$	$V_s/2$	$\frac{\sqrt{3}V}{2}(-30^\circ)$

3.3 Three Phase Half Wave Controlled Rectifier, Semi and Full Converters

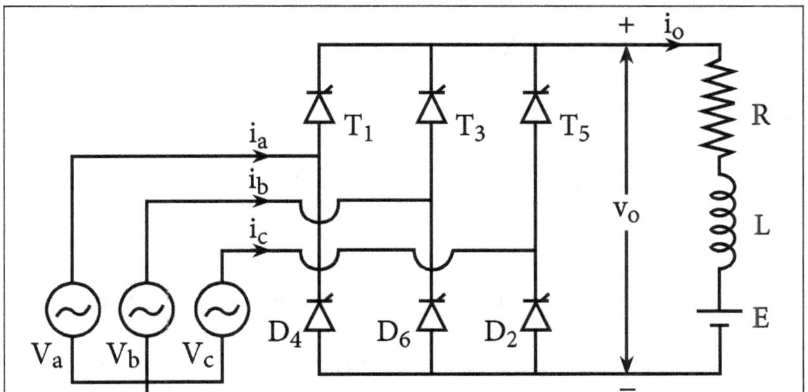

(a)

Circuit diagram of three phase half controlled converter.

The figure (a) shows the circuit diagram of the three phase half controlled converter supplying an R-L-E load. In the continuous conduction mode only one thyristor from the top group and also only one diode from bottom group conduct at a time. However, unlike fully controlled converter here are both devices from the same phase leg can conduct at the same time. Hence, there are nine conducting modes as shown in figure (b).

Table: Conduction table of three phase half controlled converter.

Mode / Volt	T_1D_2	D_2T_3	T_3D_4	D_4T_5	T_5D_6	D_6T_1	T_1D_4	T_3D_6	T_5D_2
VT_1	0	V_{ab}	V_{ab}	V_{ac}	V_{ac}	0	0	V_{ab}	V_{ac}
VD_2	0	0	V_{ac}	V_{ac}	V_{bc}	V_{bc}	V_{ac}	V_{bc}	0
VT_3	V_{ba}	0	0	V_{bc}	V_{bc}	V_{ba}	V_{ba}	0	V_{bc}
VD_4	V_{ca}	V_{ca}	0	0	V_{ba}	V_{ba}	0	V_{ba}	V_{ca}
VD_5	V_{ca}	V_{cb}	V_{cb}	0	0	V_{ca}	V_{ca}	V_{cb}	0
VT_6	V_{cb}	V_{cb}	V_{ab}	V_{ab}	0	0	V_{ab}	0	V_{cb}
V_o	V_{ac}	V_{bc}	V_{ba}	V_{ca}	V_{cb}	V_{ab}	0	0	0

Now consider the conducting and blocking state of D_2. In the blocking state, the voltage across D_2 is either v_{ac} or v_{bc}. Hence, D_2 can block only when these voltages are negative. Taking v_{bc} as the reference phasor (i.e., $v_{bc} = \sqrt{2}V_L \sin\omega t$), D_2 will block during $2\pi/3 \le \omega t \le 2\pi$ and will conduct in the interval $0 \le \omega t \le 2\pi/3$. Similarly, it can be shown that D_4 and D_6 will conduct during $2\pi/3 \le \omega t \le 4\pi/3$ and $4\pi/3 \le \omega t \le 2\pi$ respectively.

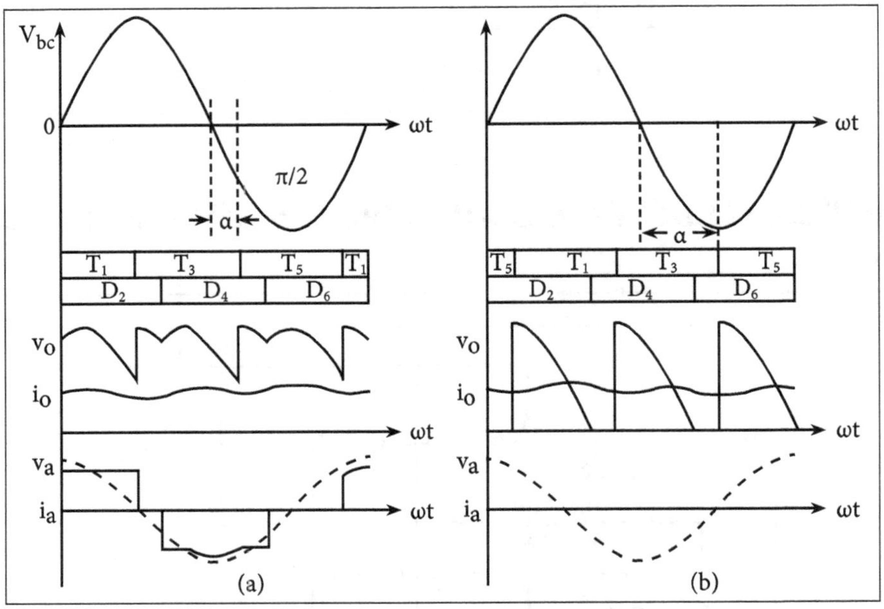

Waveform of three phase half controlled converter (a) $\alpha = \pi/6$; (b) $\alpha = \pi/2$.

Next consider conduction of T_1. The firing sequence of the thyristor is $T_1 \rightarrow T_3 \rightarrow T_5$. Before T_1 comes into conduction T_5 conducts and voltage across T_1 is $v_{ac} = \sqrt{2}V_L \sin(\omega t + \pi/3)$. If the firing angle of T_1 is α then T_1 starts conduction at $\omega t = \alpha - \pi/3$ and conducts up to $\alpha + \pi/3$. Similarly T_3 and T_5 conducts during $\alpha + \pi/3 \le \omega t \le \alpha + \pi$ and $\alpha + \pi \le \omega t \le 2\pi + \alpha - \pi/3$.

3 Phase Bridge Rectifier with R-L-E Load

A three phase fully controlled converter is obtained by replacing all the six diodes of an uncontrolled converter by six thyristors as shown in the figure. For any current to flow in the load at least one device from the top group (T_1, T_3, T_5) and one from the bottom group (T_2, T_4, T_6) must conduct. From the symmetry consideration, it can be argued that each thyristor conducts for 120° of the input cycle.

Now, the thyristors are fired in the sequence $T_1 \rightarrow T_2 \rightarrow T_3 \rightarrow T_4 \rightarrow T_5 \rightarrow T_6 \rightarrow T_1$ with 60° interval between each firing. Therefore thyristors on the same phase leg are fired at an interval of 180° and hence cannot conduct simultaneously. This leaves only six possible conduction mode for the converter in the continuous conduction mode. They are T_1T_2, T_2T_3, T_3T_4, T_4T_5, T_5T_6, T_6T_1.

Each conduction mode is of 60° duration and appears in the sequence mentioned. The conduction table of figure (b) shows voltage across different devices and the DC output voltage for each conduction interval. The phasor diagram of the line voltages are as shown in the below figure. Each of these line voltages is associated with the firing of a thyristor with the help of the conduction table. For example, the thyristor T_1 is fired at the end of T_5T_6 conduction interval.

During this period, the voltage across T_1 was v_{ac}. Therefore T_1 is fired α angle after the positive going zero crossing of v_{ac}. Similar observation can be made about other thyristors. The phasor diagram of figure also confirms that all the thyristors are fired in the correct sequence with 60° interval between each firing.

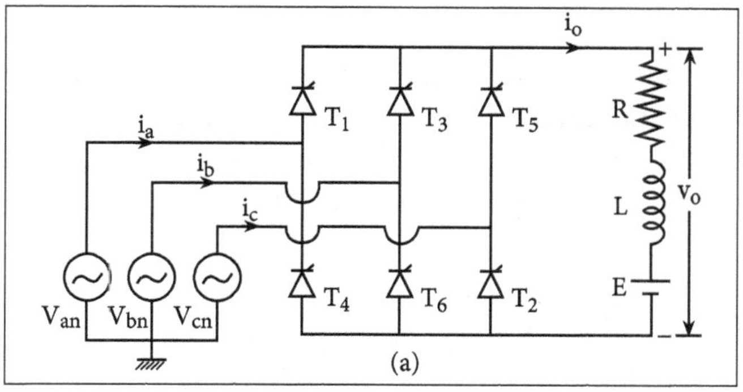

(a) Circuit Diagram.

(b) Table: Conduction table.

Device Mode	VT_1	VT_2	VT_3	VT_4	VT_5	VT_6	V_o
T_1T_2	0	0	V_{ba}	V_{ca}	V_{ca}	V_{cb}	V_{ac}
T_2T_3	V_{ab}	0	0	V_{ac}	V_{cb}	V_{cb}	V_{bc}
T_3T_4	V_{ab}	V_{ac}	0	0	V_{cb}	V_{ab}	V_{ba}
T_4T_5	V_{ac}	V_{ac}	V_{bc}	0	0	V_{ab}	V_{ca}
T_5T_6	V_{ac}	V_{bc}	V_{bc}	V_{ba}	0	0	V_{cb}
T_6T_1	0	V_{bc}	V_{ba}	V_{ba}	V_{ca}	0	V_{ab}
NONE	-	-	-	-	-	-	E

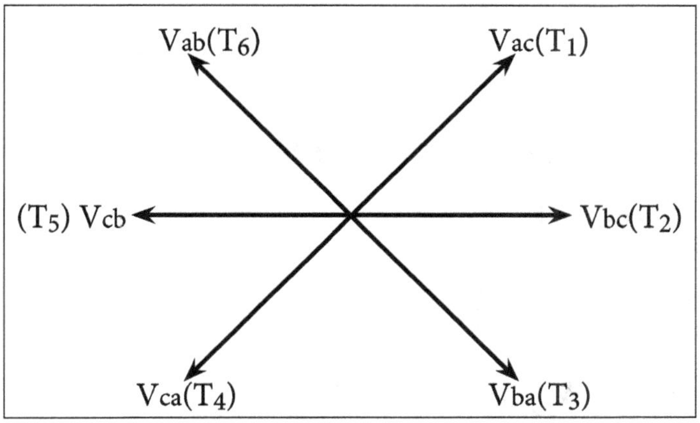

(c) Phasor Diagram of Line Voltages.

Operation of Fully Controlled Bridge Converter

To arrive at the waveforms, it is necessary to draw the conduction diagram which shows the interval of conduction for each thyristor and can be drawn with the help of the phasor diagram of the figure. If the converter firing angle is α,each thyristor is fired "α" angle after the positive going zero crossing of the line voltage with which it's firing is associated. Once the conduction diagram is drawn all other voltage waveforms can be drawn from the line voltage waveforms and from the conduction table.

Similarly, line currents can be drawn from the output current and the conduction diagram. It is clear that the output voltage and current waveforms are periodic over one sixth of the input cycle. Therefore, this converter is also termed as six pulse converter. The input current on the other hand contains only odds harmonics of the input frequency other than the triplex (3rd, 9th etc.) harmonics.

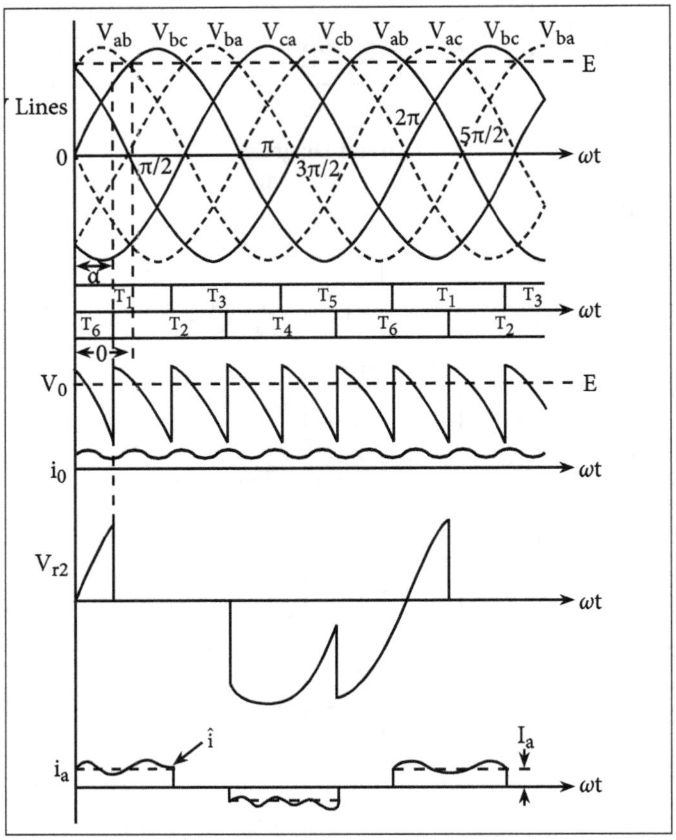

Waveforms of Three Phase Fully Controlled Converters in rectifier mode.

Analysis of Converter in the Rectifier Mode

The output voltage waveform can be written as:

$$V_o = V_o \sum_{K=1.2}^{\alpha} V_{AK} \cos 6K\omega t + \sum_{K=1.2}^{\alpha} V_{BK} \sin\ 6\ K\omega t$$

$$V_o = \frac{3}{\pi}\int_{\alpha}^{\alpha+\frac{\pi}{3}} V_o d\omega t = \frac{3\sqrt{2}}{\pi} V_L \int_{\alpha}^{\alpha+\frac{\pi}{3}} \sin\left(\omega t + \frac{\pi}{3}\right) d\omega t$$

$$= \frac{3\sqrt{2}}{\pi} V_L \cos\alpha$$

$$V_{AK} = \frac{6}{\pi}\int_{\alpha}^{\alpha+\frac{\pi}{3}} V_o \cos 6\, K\omega t\, d\omega t$$

$$= \frac{6}{\pi}\int_{\alpha}^{\alpha+\frac{\pi}{3}} \sqrt{2} V_L \sin\left(\omega t + \frac{\pi}{3}\right)\cos 6\, \omega t\, d\omega t$$

$$= \frac{3\sqrt{2}}{\pi} V_L \left[\frac{\cos(6K+1)\alpha}{6K+1} - \frac{\cos(6K-1)\alpha}{6K-1}\right]$$

$$V_{BK} = \frac{6}{\pi}\int_{\alpha}^{\alpha+\frac{\pi}{3}} V_o \sin 6\, K\omega t\, d\omega t$$

$$= \frac{6}{\pi}\int_{\alpha}^{\alpha+\frac{\pi}{3}} \sqrt{2} V_L \sin\left(\omega t + \frac{\pi}{3}\right)\sin 6\, \omega t\, d\omega t$$

$$= \frac{3\sqrt{2}}{\pi} V_L \left[\frac{\sin(6K+1)\alpha}{6K+1} - \frac{\sin(6K-1)\alpha}{6K-1}\right]$$

$$V_{oRMS} = \sqrt{\frac{3}{\pi}\int_{\alpha}^{\alpha+\frac{\pi}{3}} v_o^2 d\omega t} = V_L \left[1 + \frac{3\sqrt{3}}{4\pi}\cos 2\alpha\right]^{\frac{1}{2}}$$

3.3.1 Input Side Current Harmonics and Power Factor

Output DC Power

$$P_{O(dc)} = V_{O(dc)} \times I_{O(dc)};\ \text{i.e., } P_{dc} = V_{dc} \times I_{dc}$$

Where $V_{o(dc)} = V_{dc} =$ Average or DC value of output voltage.

$I_{o(dc)} = I_{dc} =$ Average or DC value of output current.

Output AC Power

$$P_{O(ac)} = V_{O(RMS)} \times I_{O(RMS)}$$

Efficiency of Rectification (Rectification Ratio)

$$\text{Efficiency } \eta = \frac{P_{O(dc)}}{P_{O(ac)}};\ \%\text{Efficiency } \eta = \frac{P_{O(dc)}}{P_{O(ac)}} \times 100$$

The output voltage can be considered as being composed of two components:

- The DC component, $V_{o(dc)}$ = DC or average value of output voltage.

- The AC component or the ripple component, $V_{ac} = V_{rms}$ = RMS value of all the AC ripple components.

The total RMS value of output voltage is given by:

$$V_{O(RMS)} = \sqrt{V_{O(dc)}^2 + V_{r(rms)}^2}$$

Therefore,

$$V_{ac} = V_{r(rms)} = \sqrt{V_{O(RMS)}^2 - V_{O(dc)}^2}$$

Form Factor (FF):

It is a measure of the shape of the output voltage and is given by:

$$FF = \frac{V_{O(RMS)}}{V_{O(dc)}} = \frac{\text{RMS output (load) voltage}}{\text{DC output (load) voltage}}$$

The Ripple Factor (RF):

It is a measure of the AC ripple content in the output voltage waveform. The output voltage ripple factor defined for the output voltage waveform is given by:

$$r_v = RF = \frac{V_{r(rms)}}{V_{O(dc)}} = \frac{V_{ac}}{V_{dc}}$$

$$r_v = \frac{\sqrt{V_{O(RMS)}^2 - V_{O(dc)}^2}}{V_{O(dc)}} = \sqrt{\left[\frac{V_{O(RMS)}}{V_{O(dc)}}\right]^2 - 1}$$

Therefore,

$$r_v = \sqrt{FF^2 - 1}$$

Transformer Utilization Factor (TUF):

$$TUF = \frac{P_{O(dc)}}{V_S \times I_S}$$

Where V_s = RMS value of transformer secondary output voltage (RMS supply voltage at the secondary).

I_s = RMS value of transformer secondary current (RMS line or supply current).

V_s = Supply voltage at the transformer secondary side.

i_s = Input supply current.

i_{s1} = Fundamental component of the input supply current.

I_p = Peak value of the input supply current.

Φ = Phase angle difference between (sine wave components) the fundamental components of input supply current and the input supply voltage.

Φ = Displacement angle (phase angle).

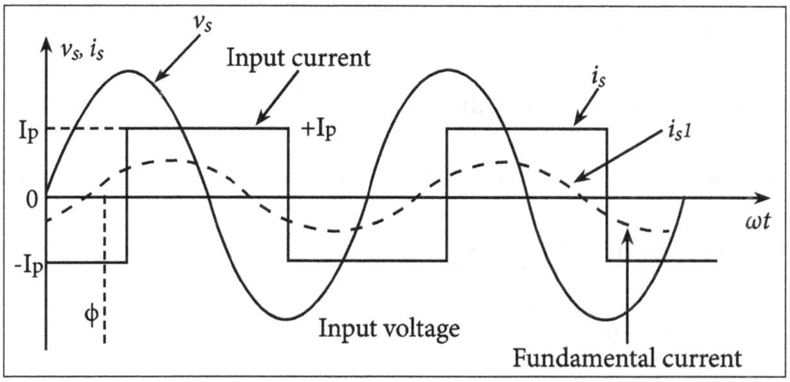

TUF characteristics.

For an RL load Φ = Displacement angle = Load impedance angle:

$$\therefore \qquad \phi = \tan^{-1}\left(\frac{\omega L}{R}\right) \text{ for an RL load}$$

Displacement Factor (DF) or Fundamental Power Factor

$$DF = \cos\phi$$

Harmonic Factor (HF) or Total Harmonic Distortion Factor (THD)

The harmonic factor is a measure of the distortion in the output waveform and is also referred to as the total harmonic distortion (THD).

$$HF = \left[\frac{I_S^2 - I_{S1}^2}{I_{S1}^2}\right]^{\frac{1}{2}} = \left[\left(\frac{I_S}{I_{S1}}\right)^2 - 1\right]^{\frac{1}{2}}$$

Where I_s = RMS value of input supply current.

I_{S1} = RMS value of fundamental component of the input supply current.

Input Power Factor (PF)

$$PF = \frac{V_S I_{S1}}{V_S I_S}\cos\phi = \frac{I_{S1}}{I_S}\cos\phi$$

The Crest Factor (CF)

$$CF = \frac{I_{S(peak)}}{I_S} = \frac{\text{Peak input supply current}}{\text{RMS input supply current}}$$

For an Ideal Controlled Rectifier

- FF = 1; which means that $V_{O(RMS)} = V_{O(dc)}$.
- Efficiency η= 100%; which means that = $P_{O(dc)} = P_{O(ac)}$.
- $V_{ac} = V_{r(rms)} = 0$; so that RF = r_v = 0 ; Ripple factor = 0 (ripple free converter).
- TUF = 1; which means that $P_{O(dc)} = V_S * I_S$.
- HF = THD = 0; which means that $I_s = I_{s1}$.
- PF = DPF = 1; which means that Φ = 0.

3.4 Dual Converters: Circulating Current Mode and Non-circulating Current Mode

Single Phase Dual Converter

We have seen in the case of a single phase full converter with inductive loads, the converter can operate in two different quadrants in the V_{dc} versus I_{dc} operating diagram.

If two single phase full converters are connected in parallel and in opposite direction (connected in back to back) across a common load, four quadrant operations are possible. Such a converter is called as a dual converter which is shown in the figure:

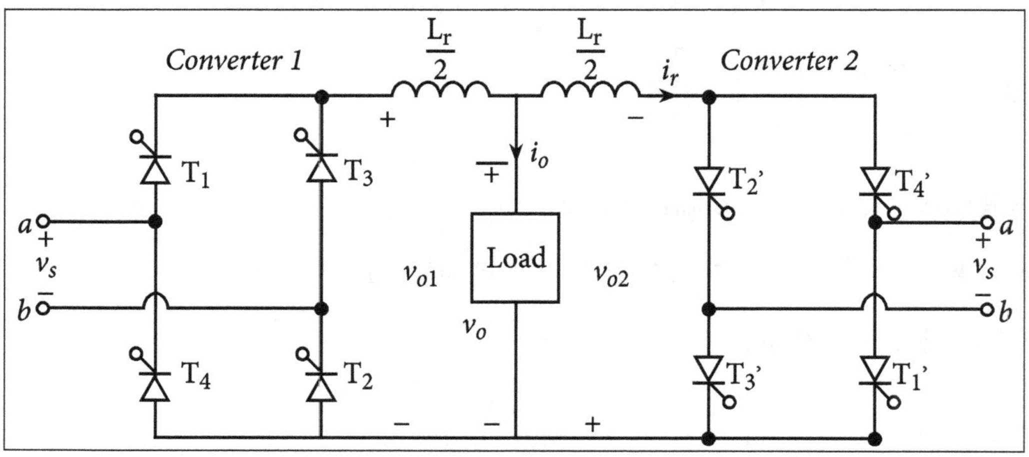

Single phase dual converter circuit.

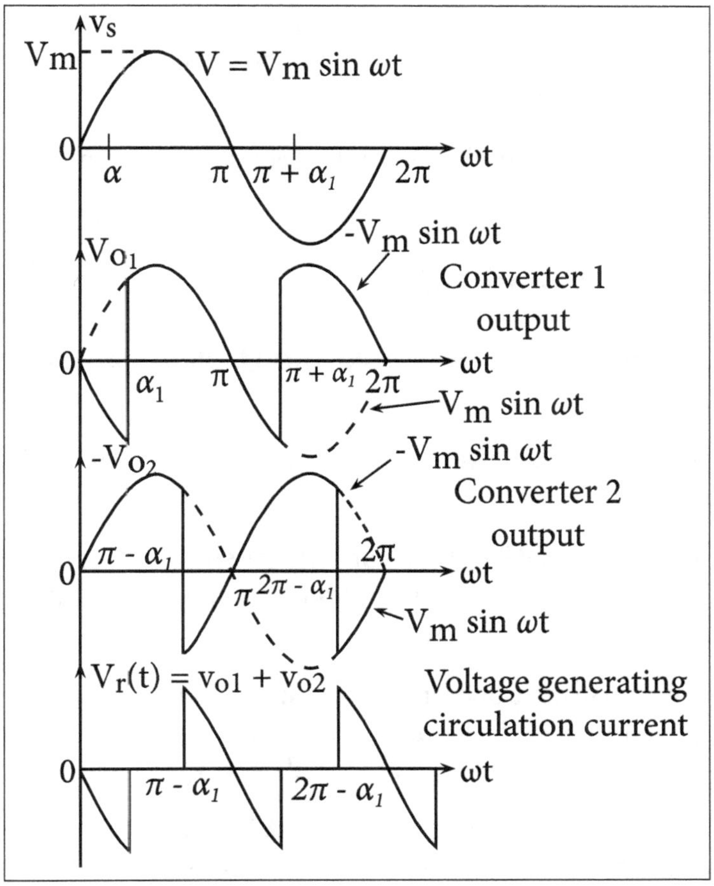

Single phase dual converter waveform.

The dual converter system will provide four quadrant operations and is normally used in high power industrial variable speed drives.

The converter 1 provides a positive DC output voltage and a positive DC load current when operated in the rectification mode.

The converter 2 provides a negative DC output voltage and a negative DC load current when operated in the rectification mode.

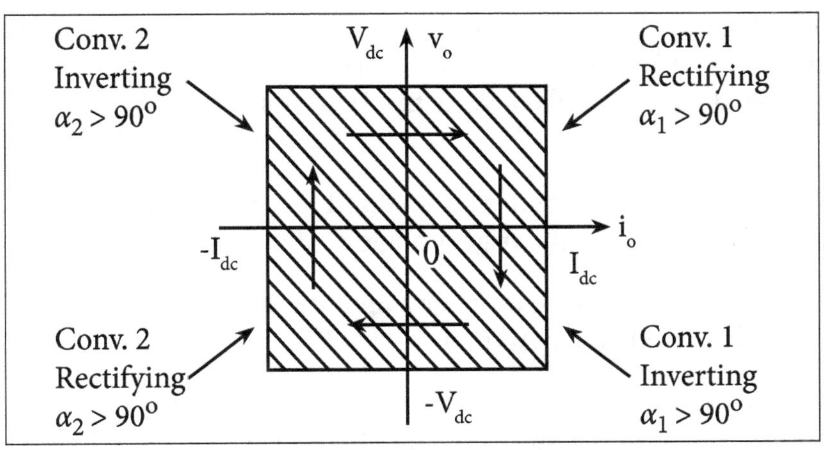

Four quadrant operation of a dual converter.

We can thus have bi-directional load current and bi-directional DC output voltage. The magnitude of output DC load voltage and the DC load current can be controlled by varying the trigger angles α_1 and α_2 of the converters 1 and 2 respectively.

There are two modes of operations possible for a dual converter system:

- Non-circulating current mode of operation (Circulating current free mode of operation).
- Circulating current mode of operation.

Non-circulating Current Mode of Operation (Circulating Current Free Mode Of Operation)

In this mode of operation, only one converter is switched ON at a time while the second converter is switched OFF.

When the converter 1 is switched ON and the gate trigger signals are released to the gates of thyristors in converter 1, we get an average output voltage across the load which can be varied by adjusting the trigger angle α_1 of the converter 1.

If α_1 is less than 90°, the converter 1 operates as a controlled rectifier and converts the input AC power into DC output power to feed the load. V_{dc} and I_{dc} are both positive and the operation occurs in the first quadrant. The average output power period, $s_c = V_{dc} \times I_{dc}$ is positive.

The power flows from the input AC supply to the load. When α_1 is increased above 90°, converter 1 operates as a line commutated inverter and V_{dc} becomes negative while I_{dc} is positive and the output power period s_c becomes negative. The power is fed back from the load circuit to the input AC source through the converter 1. The load current falls to zero when the load energy is utilized completely.

The second converter 2 is switched ON after a small delay of about 10 to 20 milliseconds to allow all the thyristors of converter 1 to turn OFF completely. The gate signals are released to the thyristor gates of converter 2 and the trigger angle α_2 is adjusted such that $\alpha_2 < 90°$ that converter 2 operates as a controlled rectifier.

The DC output voltage V_{dc} and I_{dc} are both negative and the load current flows in the reverse direction. The magnitude of V_{dc} and I_{dc} are controlled by the trigger angle α_2. The operation occurs in the third quadrant where V_{dc} and I_{dc} are both negative and output power P is positive and the converter 2 operates as a controlled rectifier and converts the AC supply power into DC output power which is fed to the load.

When we want to reverse the load current flow so that I_{dc} is positive, we have to operate converter 2 in the inverter mode by increasing the trigger angle α_2 above 90°.

When α_2 is made greater than 90°, the converter 2 operates as a line commutated inverter and the load power (load energy) is fed back to AC mains.

The current falls to zero when all the load energy is utilized and the converter 1 can be switched ON after a short delay of 10 to 20 milliseconds to ensure that the converter 2 thyristors are completely turned OFF.

The advantage of non-circulating current mode of operation is that there is no circulating current flowing between the two converters as only one converter operates and conducts at a time while the other converter is switched OFF. Hence, there is no need of the series current limiting inductor between the outputs of the two converters.

The current rating of thyristors is lower in this mode. But the disadvantage is that the load current tends to become discontinuous and the transfer characteristic becomes nonlinear. The control circuit becomes complex and the output response is sluggish as the load current reversal takes some time due to the time delay between the switching OFF of one converter and the switching ON of the other converter.

Hence, the output dynamic response is poor. Whenever a fast and frequent reversal of the load current is required, the dual converter is operates in the circulating current mode.

Circulating Current Mode of Operation

In this mode of operation both the converters 1 and 2 are switched ON and operated simultaneously and both the converters are in a state of conduction.

The average DC output voltage of converter 1 is:

$$V_{dc1} = \frac{2V_m}{\pi}\cos\alpha_1$$

The average output voltage of converter 2 is:

$$V_{dc2} = \frac{2V_m}{\pi}\cos\alpha_2$$

In the dual converter operation, one converter is operated as a controlled rectifier with α_1>90° and the second converter is operated as a line commutated inverter in the inversion mode with α_2 >90°.

$$V_{dc1} = -V_{dc2}$$

$$\frac{2V_m}{\pi}\cos\alpha_1 = \frac{-2V_m}{\pi}\cos\alpha_2 = \frac{2V_m}{\pi}(-\cos\alpha_2)$$

Therefore:

$$\cos\alpha_1 = -\cos\alpha_2 \text{ or } \cos\alpha_2 = -\cos\alpha_1 = \cos(\pi-\alpha_1)$$

Therefore:

$$\alpha_2 = (\pi-\alpha_1) \text{ or } (\alpha_1+\alpha_2) = \pi \text{ radians}$$

Which gives:

$$\alpha_2 = (\pi-\alpha_1)$$

When the trigger angle a_1 of converter 1 is set to some value, the trigger angle a_2 of the second converter is adjusted such that a_2= (180° - a_1). Hence for circulating current mode of operation where both converters conduct at the same time (a_1+a_2) = 180° so that they produce the same DC output voltage across the load.

When a_1 < 90° (say a_1=30°) the converter 1 operates as a controlled rectifier and converts the AC supply into DC output power and the average load current I_{dc} is positive. At the same time the converter 2 is switched on and operated as a line commutated inverter by adjusting the trigger angle a_2 such that a_2 = (180°-a_1), which is equal to 150°, when a_1=30°.

The converter 2 will operate in the inversion mode and feeds the load energy back to the AC supply. When we want to reverse the load current flow we have to switch the roles of the two converters.

When converter 2 is operated as a controlled rectifier by adjusting the trigger angle a_2 such that a_2 <90°, the first converter 1 is operated as a line commutated inverter by adjusting the trigger angle a_1 such that a_1>90°. The trigger angle a1 is adjusted such that a_1 = (180°- a_2) for a set value of a_2.

In the circulating current mode, a current builds up between the two converters even when the load current falls to zero. In order to limit the circulating current flowing between the two converters, we have to include current limiting reactors in series between the output terminals of the two converters.

The advantage of the circulating current mode of operation is that we can have faster reversal of load current as the two converters are in a state of conduction simultaneously.

The disadvantage of the circulating current mode of operation is that a current flows continuously in the dual converter circuit even at times when the load current is zero. Hence we should connect current limiting inductors in order to limit the peak circulating current within specified value.

3.5 AC Voltage Regulators and DC Choppers

AC Voltage Regulators

AC voltage regulators have a constant AC supply input and incorporate semiconductor switches which may vary the rms voltage impressed across the AC load. These regulators fall into the category of naturally commutating converters since their thyristor switches are commutated by the alternating supply. This converter turn-off process is termed as line commutation.

The regulator output current may be discontinuous or non-sinusoidal and as a consequence, input power factor correction and harmonic reduction are usually necessary, particularly at low output voltage levels. The feature of direction conversion of AC to AC is the absence of any intermediate energy stage, such as a DC link. Therefore, AC to AC converters are potentially more efficient but usually involve a larger number of switching devices and output is lost if the input supply is temporarily lost.

There are three basic AC regulator categories, depending on the relationship between the input supply frequency f_s which usually assume the single frequency sinusoidal and the output frequency

f_o. Without the use of transformers, the output voltage rms magnitude V_{Orms} is less than or equal to the input voltage rms magnitude V_s.

i.e., $V_{Orms} \leq V_s$

- Output frequency decreased, $f_o < f_s$.
- Output frequency increased, $f_o > f_s$.
- Output frequency fundamental supply frequency, $f_o = f_s$.

Chopper

A chopper is basically a DC to DC converter whose main function is to create a adjustable DC voltage from fixed DC voltage source through the use of semiconductors.

Types of choppers:

There are two types of choppers – AC and DC.

AC Link Chopper

In the case of an AC link chopper, first DC is converted to AC with the help of an inverter. After that, AC is stepped-up or stepped-down by a transformer, which is then converted back to DC by a diode rectifier. AC link chopper is costly, bulky and less efficient as the conversion is done in two stages.

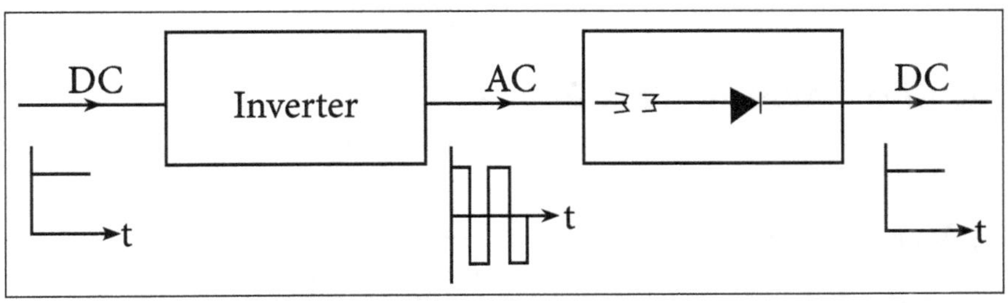

AC link chopper.

DC Chopper

A DC chopper is a static device that converts fixed DC input voltage to a variable DC output voltage directly. A chopper can be said as DC equivalent of an AC transformer, as they behave in an identical manner. This kind of chopper is more efficient, as they involve in one stage conversion.

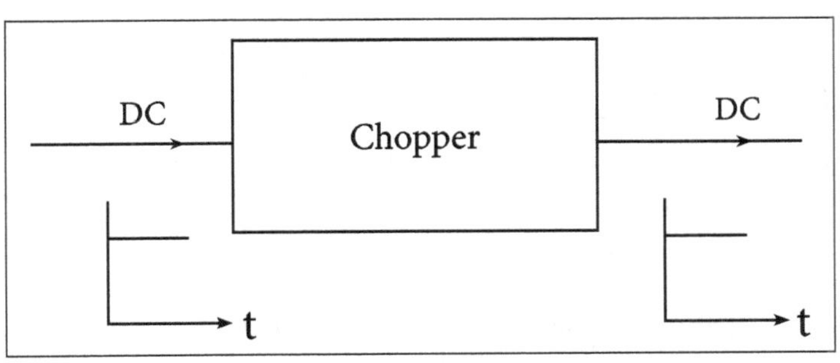

DC chopper.

Just like a transformer, a chopper can be used to step up or step down the fixed DC output voltage. Choppers are used in many applications inside various electronic equipment. A chopper system has fast response, high efficiency and a smooth control.

Principle of Operation

A chopper can be said as a high speed ON/OFF semiconductor switch. Source to load connection and disconnection from load to source happens in a rapid speed. Consider the below figure, here a chopped load voltage can be obtained from a constant DC supply, which has a magnitude Vs. Chopper is the one represented by SW inside a dotted square, which can be turned ON or OFF as desired.

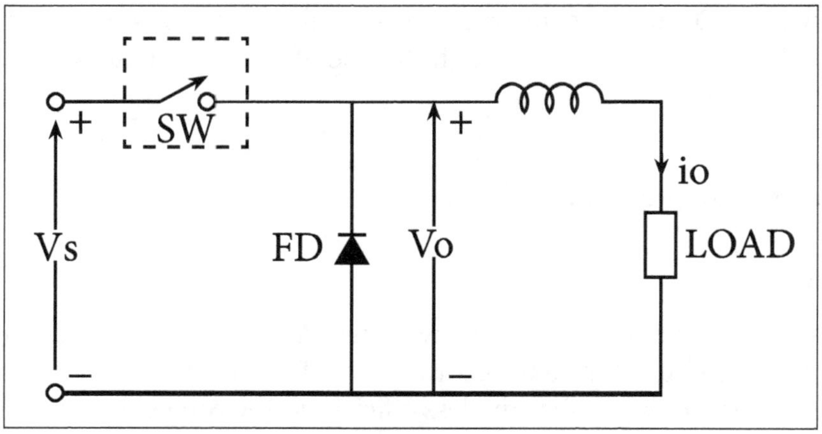

Chopper circuit.

Output Voltage and Current Waveforms

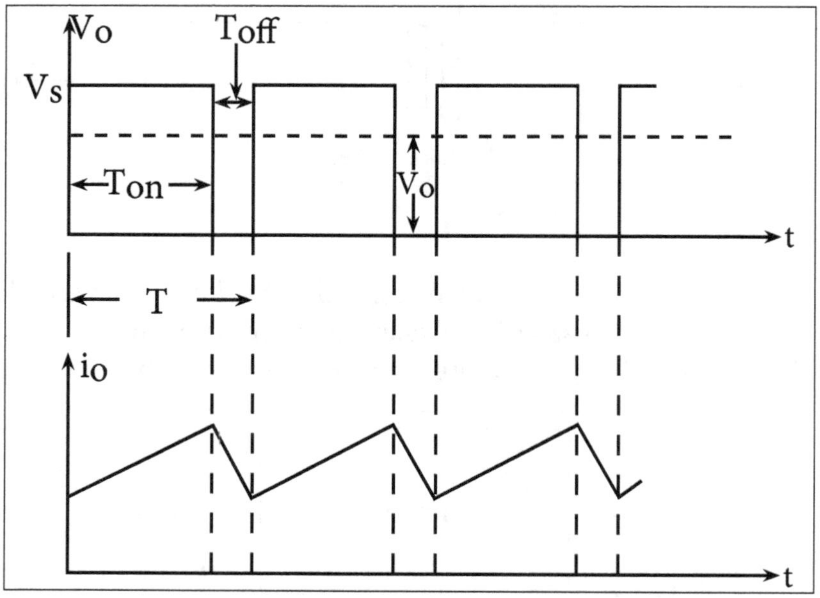

Output voltage and current waveforms.

During the time period T_{on}, the chopper is turned ON and the load voltage is equal to the source voltage Vs. During the interval T_{off}, the chopper is OFF and the load current will be flowing through the freewheeling diode FD. The load terminals are short circuited by FD and therefore, the load voltage is zero during T_{off}.

Thus, a chopped DC voltage is produced at the load terminals. We can see from the graph that, the load current is continuous. During the time period T_{on}, load current rises, but during T_{off} load current decays.

Average load voltage is given by:

$$Vo = T_{on}/(T_{on} + T_{off}) * V_s = (T_{on}/T)\,V = A\,Vs \qquad ...(1)$$

Where,

T_{on} = ON time

T_{off} = OFF time

$T = T_{on} + T_{off}$ = Chopping period

$A = T_{on}/T$ = Duty cycle

So, we know that the load voltage can be controlled by varying the duty cycle A. Equation (1) shows that the load voltage is independent of the load current and it can be written as:

$$V_o = f.T_{on}.V_s$$

f = 1/T = Chopping frequency

3.5.1 Types of AC Voltage Regulators: Single Phase Full Wave AC Voltage Controllers

Single-Phase AC Voltage Controller

The half wave AC controller uses one thyristor and one diode connected in the parallel across each other in opposite direction, that is anode of thyristor T_1 is connected to cathode of the diode D_1 and the cathode of T_1 is connected to the anode of D_1 .The output voltage across the load resistor 'R' and therefore the AC power flow to the load is controlled by varying the trigger angle 'a'.

The trigger angle or the delay angle 'a 'refers to the value of ωt or the instant at which the thyristor T_1 is triggered to turn it ON by applying a suitable gate trigger pulse between the gate and cathode lead.

The thyristor T_1 is forward biased during the positive half cycle of input AC supply. It can be triggered and made to conduct by applying a suitable gate trigger pulse only during positive half cycle of input supply. When T_1 is triggered it conducts and its load current flows through thyristor T_1, the load and through the transformer secondary winding.

By assuming T_1 as an ideal thyristor switch it can be considered as a closed switch when it is ON during the period $\omega t = \alpha$ to π radians. The output voltage across the load follows the input supply voltage when the thyristor T_1 is turned ON and when it conducts from $\omega t = \alpha$ to π radians.

When input supply voltage decreases to zero at $\omega t = \pi$ for a resistive load, the load current also falls to the value zero at $\omega t = \pi$ and hence the thyristor T_1 turns off at $\omega t = \pi$. Between the time period $\omega t = \pi$ *to* 2π, when the supply voltage reverses and becomes negative, the diode D_1 becomes forward biased and hence turns ON and conducts.

The load current flows in the opposite direction during $\omega t = \pi$ *to* 2π, radians when D_1 is ON and the output voltage follows the negative half cycle of input supply.

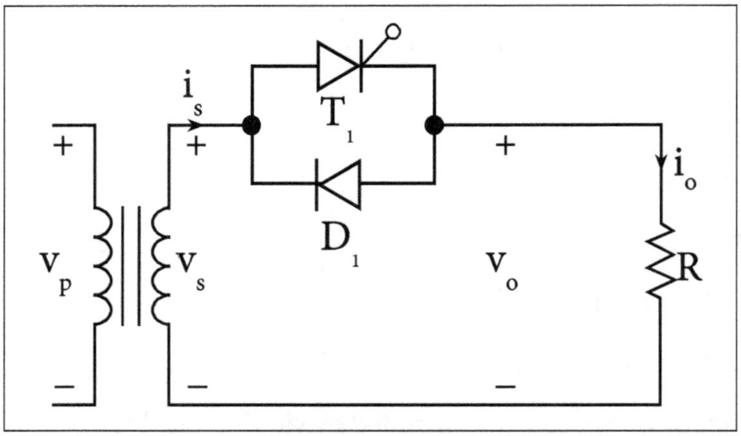

Single-Phase AC Voltage Controller.

Resistive Load and Quantitative Analysis

RMS value of output voltage is:

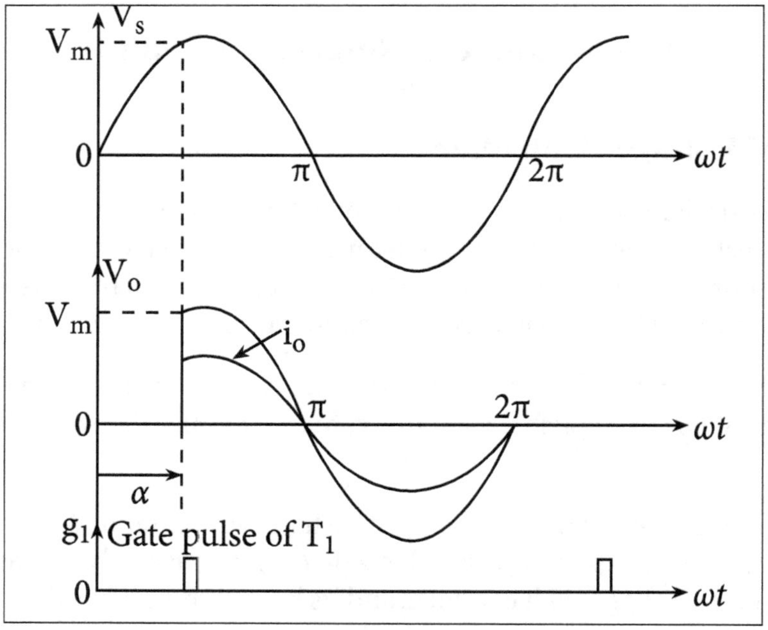

$$v_s = V_m \sin \omega t$$

$$V_S = V_{in(RMS)} = \frac{V_m}{\sqrt{2}} = \text{RMS value of secondary supply voltage.}$$

Output Load Voltage

$$v_o = v_L = 0;\ \text{for } \omega t = 0 \text{ to } \alpha$$
$$v_o = v_L = V_m \sin \omega t;\ \text{for } \omega t = \alpha \text{ to } 2\pi.$$

Output Load Current

$$i_o = i_L = \frac{v_o}{R_L} = \frac{V_m \sin \omega t}{R_L};\ \text{for } \omega t = \alpha \text{ to } 2\pi.$$
$$i_o = i_L = 0;\ \text{for } \omega t = 0 \text{ to } \alpha.$$

Input AC supply voltage across the transformer secondary winding:

RMS value of output current is,

$$I_o = \frac{U_o}{R}$$

RMS value of thyristor current is:

$$I_T = \sqrt{\frac{1}{2\pi}\int_{\alpha}^{\pi}\left(\frac{\sqrt{2}U_1 \sin \omega t}{R}\right)^2 d(\omega t)} = \frac{U_1}{R}\sqrt{\frac{1}{2}\left(1 - \frac{\alpha}{\pi} + \frac{\sin 2\alpha}{2\pi}\right)}$$

Power factor of the circuit:

$$\lambda = \frac{P}{S} = \frac{U_o I_o}{U_1 I_o} = \frac{U_o}{U_1} = \sqrt{\frac{1}{2\pi}\sin 2\alpha + \frac{\pi - \alpha}{\pi}}$$

Single-Phase Full Wave AC Voltage Controller

The basic power circuit of a single-phase ac-ac voltage controller, as shown in figure (a), comprises a pair of SCR's connected back-to-back between the AC supply and the load. This connection provides a bidirectional full-wave symmetrical control and the SCR pair can be replaced by a triac for low-power applications. Alternate arrangements are shown in figure (c). It has two diodes and two SCR's to provide a common cathode connection for simplifying the gating circuit without requiring isolation and in figure(d) with one SCR and four diodes to reduce the device cost but with increased device conduction loss.

An SCR and diode combination known as thyrode controller, as shown in figure (e), provides a uni-directional half-wave asymmetrical voltage control with device economy, but introduces the DC component and more harmonics and thus it is not so practical to use except for very low-power-heating load.

With phase control, the switches conduct the load current for a chosen period of each input cycle of voltage and with on/off control, the switches connect the load either for a few cycles of input voltage and disconnect it for the next few cycles or the switches are turned on and off several times within the alternate half cycles of input voltage.

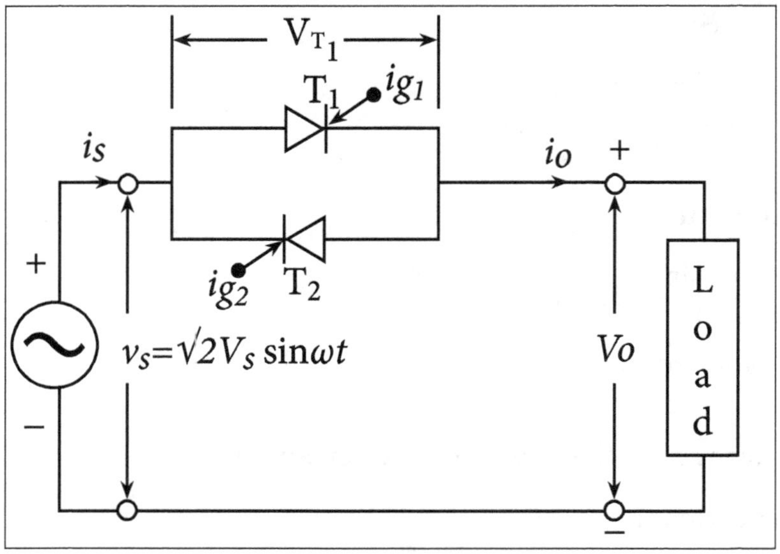

(a) Full wave, two SCR's in inverse-parallel.

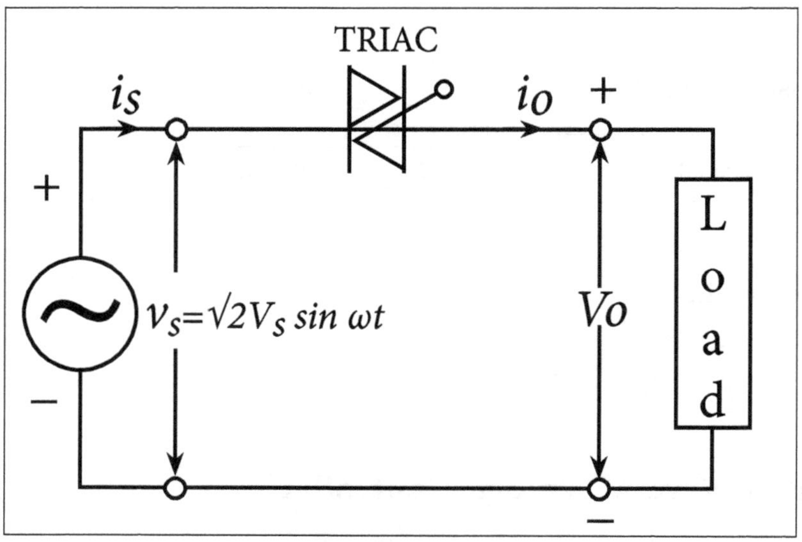

(b) Full wave with triac.

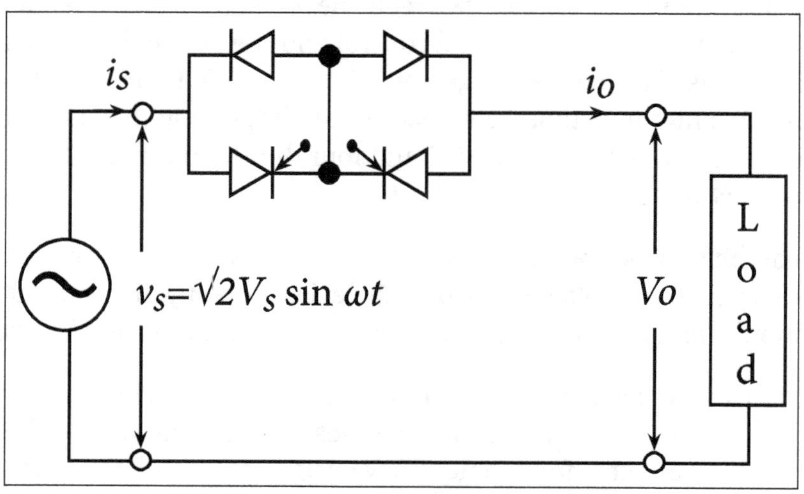

(c) Full wave with two SCR's and two diodes.

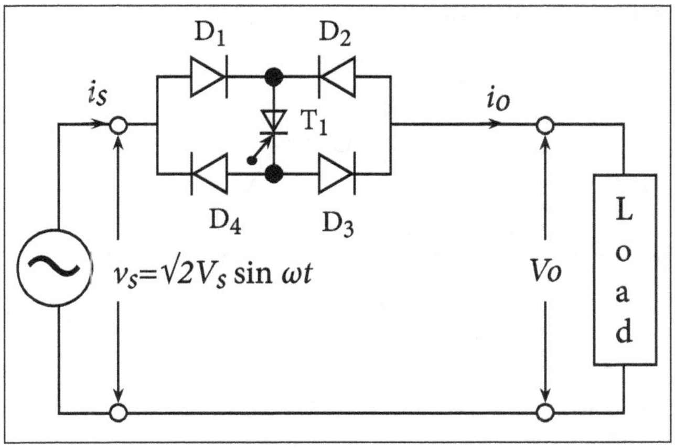

(d) Full wave with four diodes and one SCR.

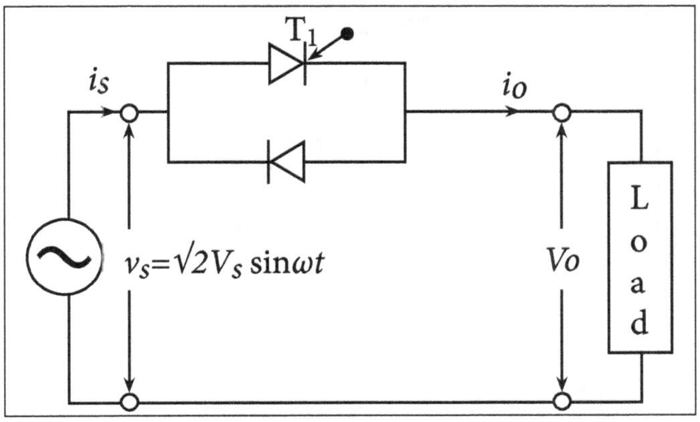

(e) Half wave with one SCR and one diode in antiparallel.

Phase-Controlled Single-phase AC Voltage Controller

For a full-wave symmetrical phase control, the SCR's T_1 and T_2 in figure (a) are gated at a and $\pi + \alpha$, respectively, from the zero crossing of the input voltage; by varying a, the power flow to the load is controlled through voltage control in alternate half cycles. As long as one SCR is carrying current, the other SCR remains reverse-biased by the voltage drop across the conducting SCR. The principle of operation in each half cycle is similar to that of the controlled half-wave rectifier and one can use the same approach for analysis of the circuit.

Operation with R-load

Figure (f) shows the typical voltage and current waveforms for the single-phase bidirectional phase-controlled AC voltage controller of figure (a) with a resistive load. The output voltage and current waveforms have half-wave symmetry and so no DC component.

If $v_s = \sqrt{2}V_s \sin \omega t$ is the source voltage, the rms output voltage with T1 triggered at α can be found from the half-wave symmetry as:

$$V_o = \left[\frac{1}{\pi}\int_{\alpha}^{\pi} 2V_s^2 \sin^2 \omega t d(\omega t)\right]^{\frac{1}{2}} = V_s\left[1-\frac{\alpha}{\pi}+\frac{\sin 2\alpha}{2\pi}\right]^{\frac{1}{2}}$$

Note that V_o can be varied from V_s to 0 by varying α from 0 to π.

The rms value of load current, $I_o = V_o / R$.

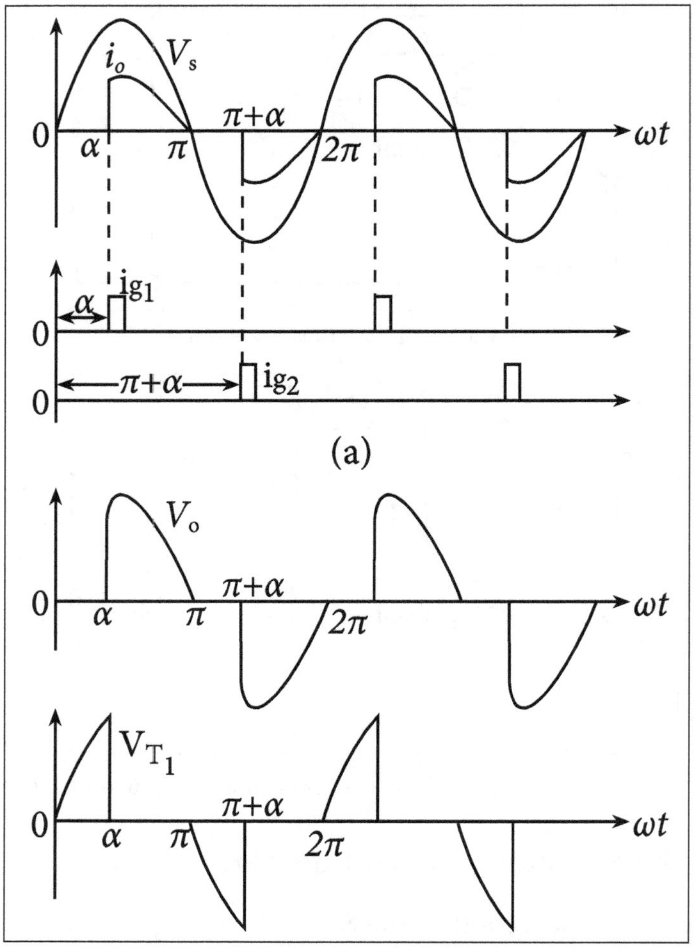

(f) Waveforms for single phase AC full wave voltage controller with R-load.

The input power factor $= \dfrac{P_o}{VA} = \dfrac{V_o}{V_s} = \left[1 - \dfrac{\alpha}{\pi} + \dfrac{\sin 2\alpha}{2\pi}\right]^{\frac{1}{2}}$

The average SCR current, $I_{A,SCR} = \dfrac{1}{2\pi R}\int_{\alpha}^{\pi} \sqrt{2}V_s \sin\omega t d(\omega t)$

Since each SCR carries half the line current, the rms current in each SCR is:

$$I_{o,SCR} = \frac{I_o}{\sqrt{2}}$$

Operation with RL Load: Figure (g) shows the voltage and current waveforms for the controller in figure (a) with RL load. Due to the inductance, the current carried by the SCR T_1 may not fall to zero at ωt = π when the input voltage goes negative and may continue till ωt = β, the extinction angle, as shown. The conduction angle, θ = β - α of the SCR depends on the firing delay angle a and the load impedance angle θ.

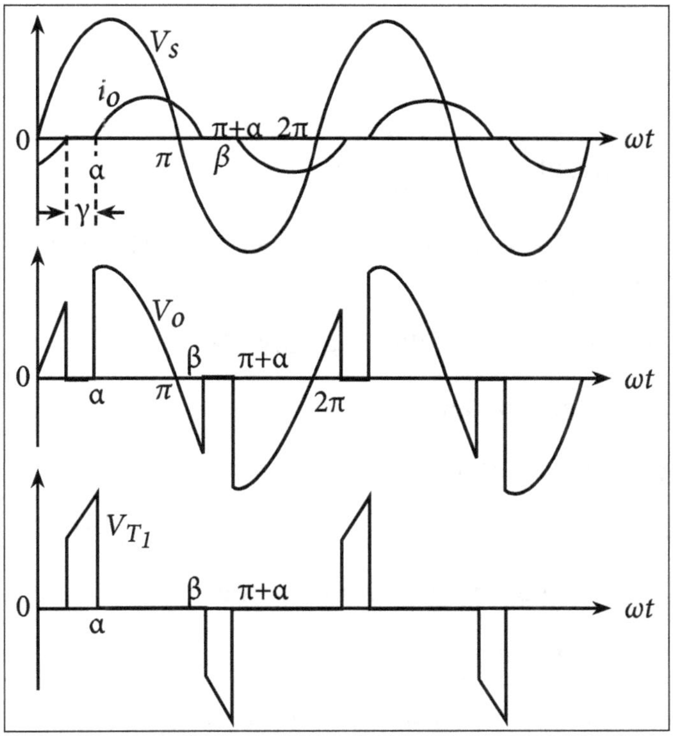

(g) Typical waveforms of single phase AC voltage controller with an RL load.

The expression for the load current $I_0(\omega t)$, when conducting from α to β can be derived in the same way as that used for a phase controlled rectifier in a discontinuous mode by solving the relevant Kirchhoff's voltage equation:

$$i_o(\omega t) = \frac{\sqrt{2}V}{Z}\left[\sin(\omega t - \phi) - \sin(\alpha - \phi)e^{(\alpha - \omega t)/\tan\phi}\right],\ \alpha < \omega t < \beta$$

Where,

$$Z = \left(R^2 + \omega^2 L^2\right)^{\frac{1}{2}}$$

The rms output voltage:

$$V_o = \left[\frac{1}{\pi}\int_{\alpha}^{\beta} 2V_s^2 \sin^2 \omega t d(\omega t)\right]^{\frac{1}{2}}$$

$$= \frac{V_s}{\pi}\left[\beta - \alpha + \frac{\sin 2\alpha}{2} - \frac{\sin 2\beta}{2}\right]^{\frac{1}{2}}$$

The rms SCR current:

$$I_{O,SCR} = \left[\frac{1}{2\pi}\int_{\alpha}^{\beta} i_o^2 d(\omega t)\right]^{\frac{1}{2}}$$

The rms load current:

$$I_o = \sqrt{2}\, I_{o,SCR}$$

The average value of SCR current, $I_{A,SCR} = \frac{1}{2\pi}\int_{\alpha}^{\beta} i_o d(\omega t)$

3.5.2 Single Phase Transformer Tap Changers: Multistep Transformer Tap Changer

Tapping are also provided for one of the following purposes:

- For varying the secondary voltage.
- For maintaining the secondary voltage constant with a varying primary voltage.
- For providing an auxiliary secondary voltage for a special purpose, such as lighting.
- For providing a low voltage for starting rotating machines.
- For providing a neutral point, e.g. for earthing.

The principal tapping is one to which the rating of the winding is related. A positive tapping means more and a negative tapping implies less turns than those of the principal tap. Tap changing may be achieved in one of the three conditions, viz:

- Voltage variation with constant flux and constant voltage turn.
- With varying flux.
- A mix of (i) and (ii).

The percentage tapping range is same as the voltage variation.

The taps may be placed on the primary or secondary side which partly depends on construction. If tappings are near the line ends, fewer bushing insulators are required. If the tappings are placed near the neutral ends, the phase-to-phase insulation conditions are eased.

No-load Tap Changing

The cheapest method of changing the turn's ratio of a transformer is the use of off-circuit tap changer.

As the name indicates, it is required to de-energize the transformer before changing the tap. A simple no-load tap changer is shown in figure.

It has eight studs marked one to eight. The winding is tapped at eight points. The face plate carrying the suitable studs can be mounted at a convenient place on the transformer such as upper yoke or located near the tapped positions on the windings. The movable contact arm A may be rotated by hand wheel mounted externally on the tank.

If the winding is tapped at 2% intervals, then as the rotatable arm A is moved over to studs 1, 2; 2,

3;......6, 7; 7, 8 the winding in circuit reduces progressively by it from 100% with arm at studs (1, 2) to 88% at studs (7, 8).

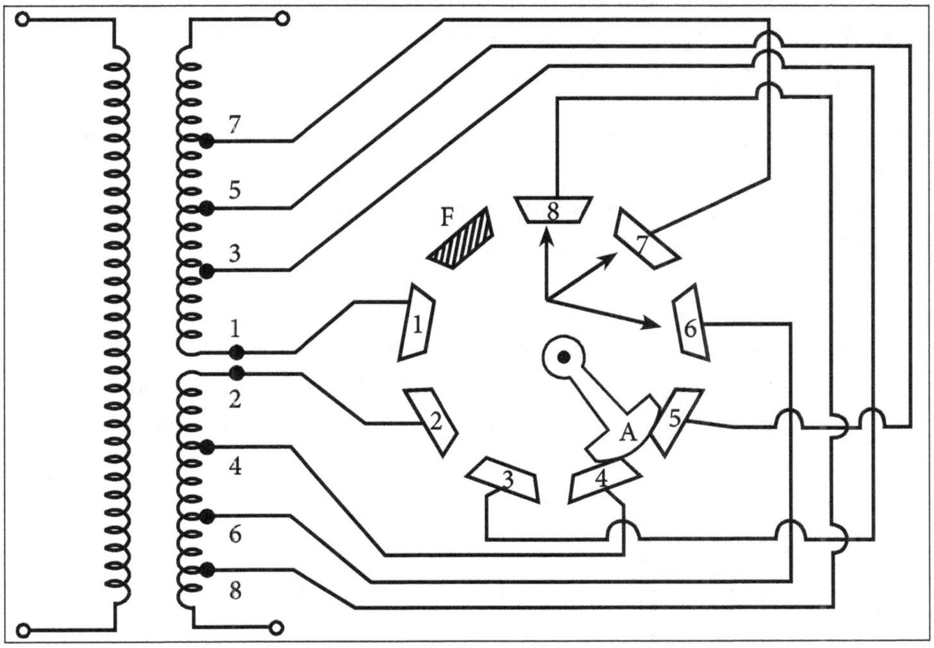

No-load tap changer.

The stop F which fixes the final position of the arm A prevents further anticlockwise rotation so that stud 1 and 8 cannot be bridged by the arm. Adjustment of tap setting is carried out with transformer de-energized.

For example, for 94% tap the arm is brought in position to bridge studs 4 and 5. The transformer can then be switched ON.

To prevent unauthorized operation of an off-circuit tap changer, a mechanical lock is provided. Further, to prevent inadvertent operation, an electromagnetic latching device or micro switch is provided to open the circuit breaker so as to de-energize the transformer as soon as the tap changer handle is moved, well before the contact of the arm with the stud (with which it was in contact) opens.

On-load Tap Changing

On-load tap changers are used to change the turn ratio of transformer to regulate system voltage while the transformer is delivering load. With the introduction of on-load tap changer, the operating efficiency of electrical system gets considerably improved.

Nowadays, almost all the large power transformers are fitted with on-load tap changer. During the operation of an on-load tap changer the main circuit should not be opened to prevent (dangerous) sparking and no part of the tapped winding should get short-circuited.

All the forms of on-load tap changing circuits are provided with impedance, which is introduced to limit short circuit current during the tap changing operation. The impedance can either be a resistor or center-tapped reactor.

The on-load tap changers can in general be classified as resistor or reactor type. In modern designs, the current limiting is almost invariably carried out by a pair of resistors. On-load tap changing gear with resistor transition, in which one winding tap is changed over for each operating position, is depicted in the figure. The figure also shows the sequence of operations during the transition from one tap to the next (in this case from tap 4 to tap 5).

Back-up main contractors are provided which short-circuit the resistor for normal operation. To ensure that the transition once started gets completed, energy storage is provided which acts even if the auxiliary power supply happens to fail. In resistor-aided tap changing, the current break is made easier by the fact that the short-circuit resistor causes the current to be opened to have unity power factor.

On-load tap changer control gear can be from simple push-button initiation to complex automatic control of several transformers operating in parallel. The aim is to maintain a given voltage level within a specified tolerance or to raise it with load to compensate for the transmission line voltage drop.

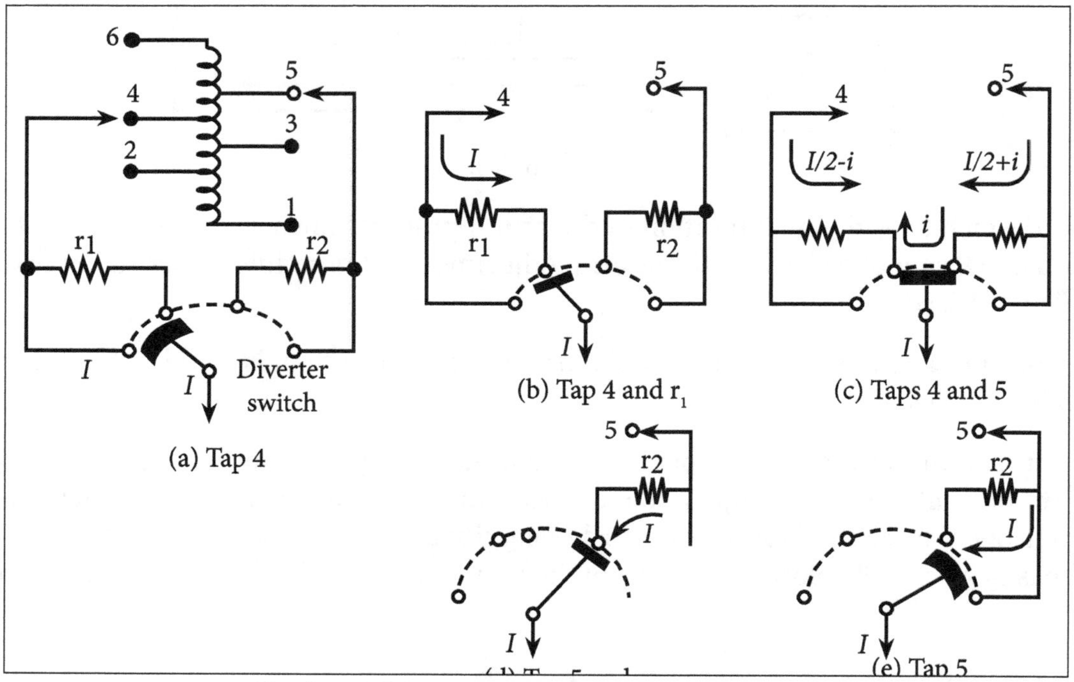

Simple switching sequence for on-load taps changing.

The main components are an automatic voltage regulator, a time delay relay and compounding elements. The time delay prevents unwanted initiation of a tap change by a small transient voltage fluctuation. It may be set for a delay up to 1 min.

Effect of Source Inductance on Phase Controlled AC-DC Converters

The presence of source inductance means that a commutation of load current from one thyristor to the next as they are triggered with a firing angle α cannot be instantaneous. This source inductance (L_s) is invariably because of the inductance of the supply lines and the leakage inductance of the input transformer.

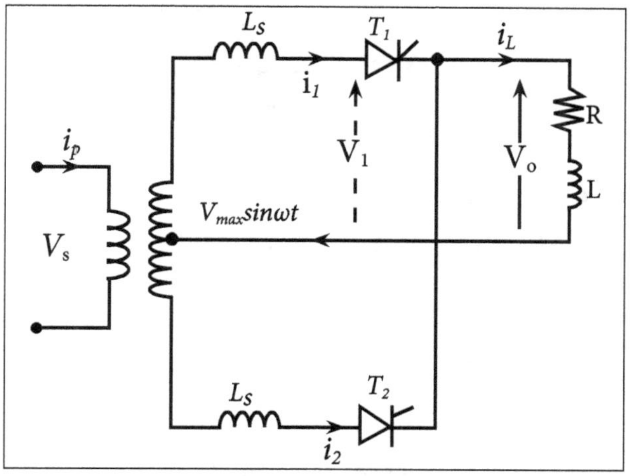

Single-phase Center Tapped Converter with Source inductance.

For this circuit, the overlap of conduction for the duration μ sec makes the output voltage zero during this period.

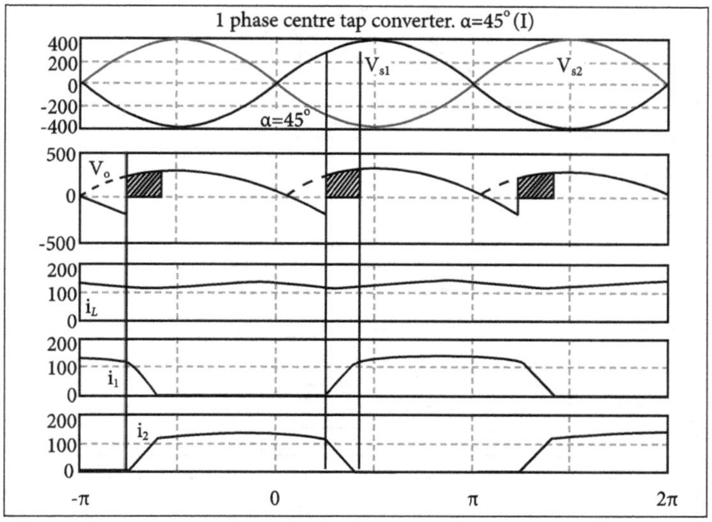

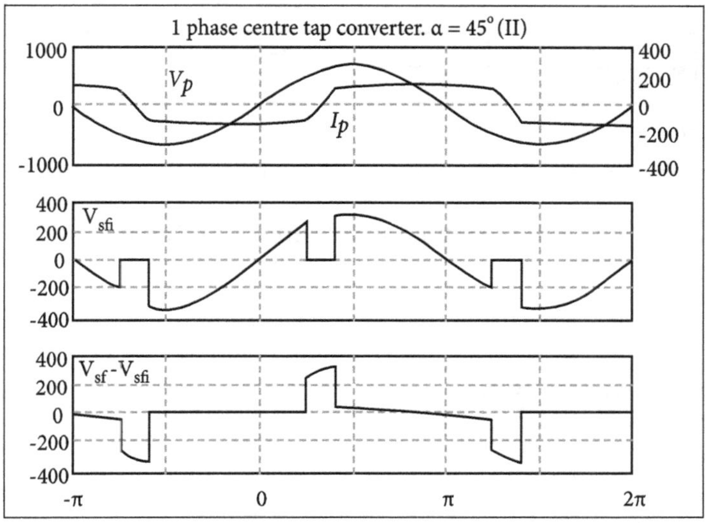

Waveforms for the converter, commutation overlap and commutation notches, for α = 45°.

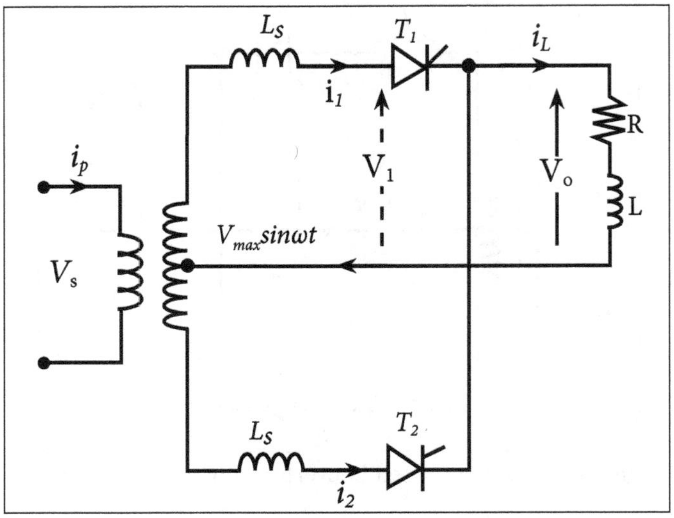

Circuit diagram.

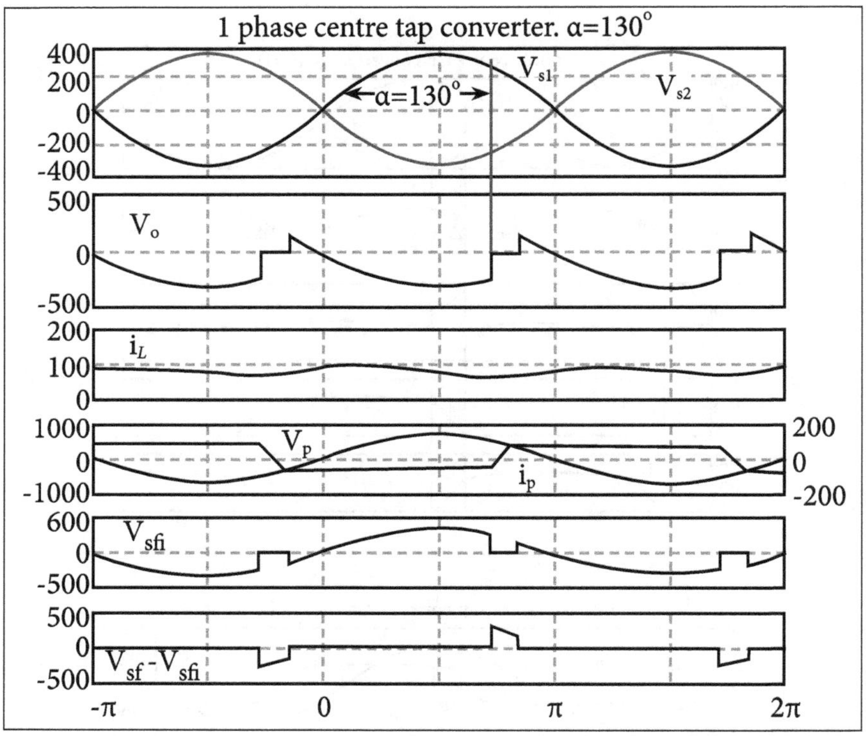

Waveforms for the converter, commutation overlap and commutation notches, for α = 130°.

The part of the input voltage waveform which is missing from the output voltage is given by:

$$V_{max} \sin \omega t = L_s \frac{di}{dt}$$

$$\int_{\alpha}^{\alpha+\mu} V_{max} \sin(\omega t) d(\omega t) = \omega L_s \int_{0}^{I_d} di$$

Where,

V_{max} is the peak of the input AC voltage.

I_d is the DC level of the output (load) current.

Note that in each half cycle of the above waveform, the input current through the incoming thyristor rises from zero to I_d in time μ/ω, starting from the instant of firing. Thus, integrating equation is given by,

$$-V_{max}\left[\cos(\alpha+\mu)-\cos\alpha\right]=\omega L_s I_d$$

$$\cos(\alpha+\mu)=\cos\alpha-\frac{\omega L_s}{V_{max}}I_d$$

The commutation or overlap angle μ can be found from the above expression. It increases with load current Id at any firing angle.

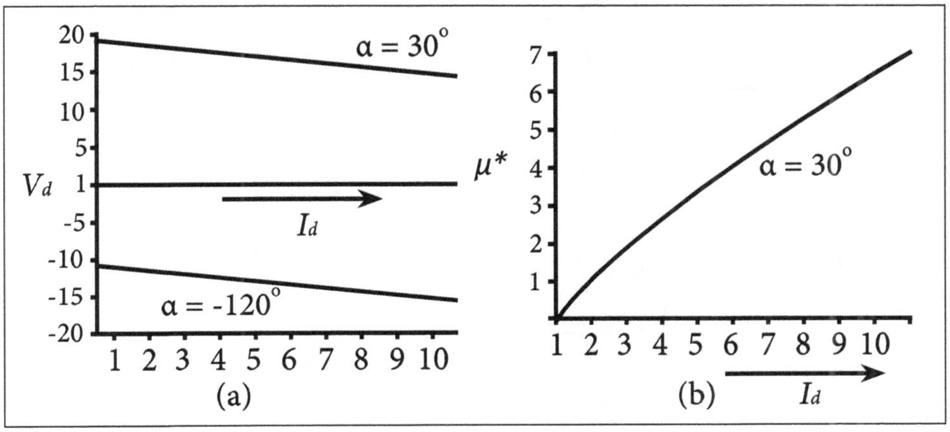

Variation of V_d and commutation angle μ with I_d and firing angle α.

The output DC voltage is then given by:

$$V_d=\frac{1}{\pi}\int_{\alpha}^{\pi+\alpha}V_{max}\sin(\omega t)d(\omega t)-\frac{1}{\pi}\int_{\alpha}^{\alpha+\mu}V_{max}\sin(\omega t)d(\omega t)$$

$$=\frac{2V_{max}}{\pi}\cos\alpha-\frac{\omega L_s}{\pi}I_d$$

3.5.3 Three Phase AC Voltage Regulators

AC Voltage Regulators

The torque developed by the 3-phase Induction motor can be expressed as:

$$T=\frac{60}{2\pi N_S}\frac{3V_1^2R_2'/S}{\left(R_1+\frac{R_2}{S}\right)^2+(X_1+X_2')^2}$$

$V_1 \rightarrow$ stator voltage (volts)

$R_1 \rightarrow$ stator resistance (ohms)

R_2'– rotor resistance (ohms)

X_2' – rotor reactance (ohms)

X_1 – stator reactance (ohms)

N_s – Synchronous speed (rpm)

S – slip

$$N_S \propto V_1^2 \quad \dots 1 \qquad \dots(1)$$

From equation (1), it is clear that the speed of the motor can be varied by varying the stator voltage.

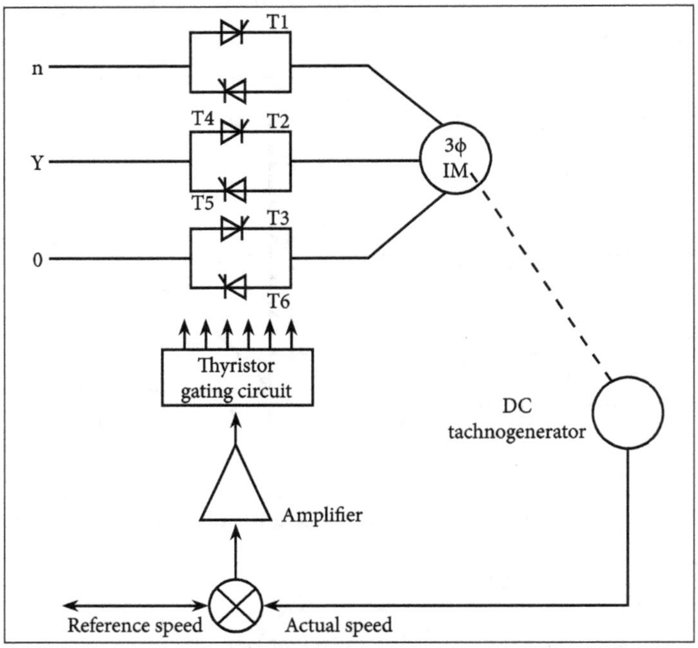

Closed loop control of IM by using stator voltage controller.

Stator voltage controller consists of a power circuit, 3ϕ Induction motor, tacho generator and control circuits. This tacho generator generates a voltage which is proportional to the motor speed. The generated voltage is equal to the actual speed of the motor. This voltage is compared with DC reference voltage.

The difference between these two voltages is compared in comparator and output voltage is error signal voltage. This voltage is amplified by amplifier and fed to the thyristor gating circuits. This controls the thyristor firing angles and thereby changes the terminal voltage and hence the motor speed changes. If the reference speed is greater than actual speed, the conduction periods of the SCRs are increased.

The increased stator voltage allows the development of an increased motor torque and hence the motor speed also increases. If the actual speed is greater than the reference speed, the conduction angle of SCRs is reduced and the motor torque decreases as well as reduces the motor speed.

When the motor speed is equal to the reference speed, the conduction angles are just sufficient to allow the development of motor torque. This motor torque is equal to the load torque. In a high-gain feedback system, the desired speed can be maintained and there is no necessity for the motor to have flat speed-torque characteristics since the output speed is determined by the reference signal rather than the open loop characteristics of the motor.

3 Phase AC Voltage Controller Fed 3 Phase Induction Motor

In 3 phase induction motor, torque T is proportional to the square of the stator voltage suppose reduction in the supply voltage will reduce the motor torque and hence the speed of the drive. In addition the motor terminal voltage is reduced to $K < 1$, where KC_1 then the motor torque is given by:

$$T_e = \frac{3}{\omega_s} \frac{(KV_1)^2}{(r_1 + r')^2 + (X_1 + X_2)^2} . r_2'$$

A 3 phase AC voltage regulator is normally employed for the purpose of varying the voltage applied to the stator of induction motor and also to achieve a speed control.

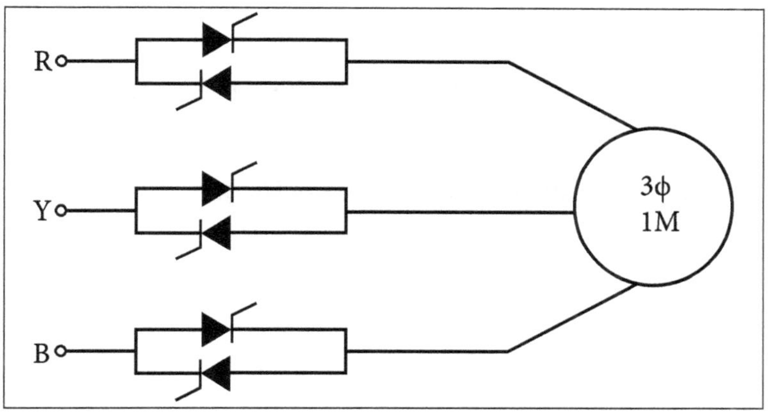

Phase AC voltage controller fed 3 phase induction motor.

In consists of 6 thyristors [$T_1 - T_6$] connected in each phase of stator terminals in anti-parallel connection of two thyristors to each phases. By introducing the firing angle of the thyristors connected in anti-parallel in each phase, the rms value of the stator voltage can be regulated. By varying the firing angle of each thyristors used in AC regulator we can vary the voltage applied to the stator terminal of 3 phase induction motor.

Advantages and disadvantages of AC voltage controller fed three phase induction motor drive when compared to inverter fed driver.

Advantages:

- It is compact in size.
- It has simple control circuit.
- Fast response time.
- It is economical as compared to other methods of speed control.

Disadvantages:

- It has less efficiency.
- High harmonics content.
- Voltage and current waves are severely distorted.
- Poor performance in running condition.
- High resistance losses.
- At low speed current drawn is more.

Problems

1. A 3-phase, star connected, 50 Hz, 4-pole induction motor has the following parameters in ohms per phase referred to the stator: $R_1 = R_2 = 0.034$ and $X_1 = X_2 = 0.18$. The motor is controlled by the variable frequency control with a constant (V/f). Let us determine the following for an operating frequency of 15 Hz:

- The breakdown torque as a ratio of its value at the rated frequency for motoring and braking.
- The starting torque and rotor current in terms of their valued at the rated frequency.

Solution:

Given:

$$R_1 = R_2 = 0.034$$

$$X_1 = X_2 = 0.18$$

Frequency - 15 Hz

$$K = 15/50 = 0.3$$

The ratio of breakdown torques for k = 0.3 and k = 1 is expressed as:

$$\frac{T_{max}(K=0.3)}{T_{max}(K=1)} = \frac{0.034 \pm \sqrt{(0.034)^2 + (0.36)^2}}{\frac{0.034}{0.3} \pm \sqrt{\left(\frac{0.034}{0.3}\right)^2 + (0.36)^2}}$$

For motoring:

$$\frac{T_{max}(K=0.3)}{T_{max}(K=1)} = 0.806$$

For braking:

$$\frac{T_{max}(K=0.3)}{T_{max}(K=1)}=1.24.$$

Substituting S=1, in the torque equation:

$$T_s=\frac{3}{\omega_s}\left[\frac{V_{rated}^2(R_2/K)}{\left(\frac{R_1+R_2}{K}\right)^2+(X_1+X_2)^2}\right]$$

The ratio of starting torques for k = 0.3 and k = -0.1

$$\frac{T_s(K=0.3)}{T_s(K=1)}=\frac{\frac{(0.034/0.3)}{(0.068/0.3)^2+(0.36)^2}}{\frac{0.034}{(0.068)^2+(0.36)^2}}=2.472$$

The starting rotor current is given by:

$$1_2=\frac{V_{rated}}{\sqrt{\left(\frac{R_1+R_2}{K}\right)^2+(X_1+X_2)^2}}$$

The ratio of starting currents for k = 0.3 and k = 1 is:

$$\frac{I_2(K=0.3)}{I_2(K=1)}=\frac{\sqrt{(0.068)^2+(0.36)^2}}{\sqrt{\left(\frac{0.068}{0.3}\right)^2+(0.36)^2}}=0.86$$

The above calculations of ratios of starting torques and starting rotor currents show that the constant (v/f) control provides a high starting torque with reduced current.

2. A three-phase, 460 V, 50 Hz, 4-pole, 1420 rpm, star-connected induction motor has the following parameters per phase referred to the stator: R_1= 0.66 Ω, R_2 = 0.38 Ω, X_1 =X_2= 1.14 Ω, X_m = 32 Ω. The motor is controlled by a variable frequency control at a constant flux of rated value. Let us determine the following:

- The motor speed and the stator current at half the rated torque and 25 Hz.
- By assuming the speed-torque curves to be straight lines, solve for part (a), for $S < S_m$.
- The frequency, the stator current, and voltage at a rated braking torque and 1200 rpm.

Solution:

Given:

$$R_1 = 0.66\ \Omega$$

$$R_2 = 0.38\ \Omega$$

$$X_1 = X_2 = 1.14\ \Omega$$

$$X_m = 32\ \Omega$$

$$N_S = \frac{120\ f}{P} = \frac{120 \times 50}{4} = 1500$$

Synchronous speed, N_s = 120f/P

$$\omega_s = \frac{1500 \times 2\pi}{60} = 157.08\ \text{rad/s}\ \text{and} \therefore S = \frac{1500 - 1420}{1500} = 0.05$$

Rotor impedance:

$$Z_2 = \frac{R_2}{S} + j \times 2 = \frac{0.38}{0.05} + j\,1.14 = 7.6 + j1.14 = 7.69\ \angle 8.53°.$$

Stator impedance:

$$Z_1 = 0.66 + j1.14 = 1.32\ \angle 59.93°$$

Machine impedance:

$$Z_{in} = Z_i + \frac{Z_2 Z_m}{Z_2 + Z_m}$$

$$= (0.66 + j\,1.14) + \frac{(7.6 + j1.14)(j32)}{(7.6 + j\,1.14 + j32)} = 6.91 + j3.01 = 7.54\ \angle 23.54°\ \Omega$$

Stator current:

$$I_1 = \frac{400/\sqrt{3}}{4} = 66.4\ \text{A}$$

$$I_2 = \left|\frac{Z_m}{Z_m + Z_2}\right| I_1 = \left|\frac{j\,32}{j\,32 + 7.6 + j\,1.14}\right| 66.4 = \left(\frac{32}{34}\right)(66.4) = 62.49\ \text{A}$$

$$E = I_2 |Z_2| = 62.49 \times 7.69 = 480.55\ \text{V}$$

$$\text{Rated torque} = \frac{3}{\omega_s} I_2^2 \frac{R_2}{S} = \frac{3}{157.08} \times (62.49)^2 \times \frac{0.38}{0.05} = 566.81\ \text{N} - \text{m}$$

At 25Hz:

Substituting the known values in torque equation:

$$\frac{566.81}{2}=\frac{3}{157.08}\left[\frac{(480.55)^2\times(0.38/0.5S)}{(0.38)^2/(0.5S)^2+(1.14)^2}\right]=S=0.0535$$

$$\omega_r = K\omega_s(1-s)$$

$$N = KN_s(1-s)$$

At 25 Hz, $E = 0.5\times 480.55 = 240.28\ V$

$$Z_2 = \frac{R_2}{S}+j\,KX_2 = \frac{0.38}{0.0535}+j0.5\times1.14 = 7.102+j\,0.57 = 7.125\ \angle 4.59°$$

Now taking E as reference vector:

$$\bar{I}_2 = \frac{E}{Z_2} = \frac{240.28}{7.125\ \angle 4.59°} = 33.72\angle -4.59°\ A$$

$$\bar{I}_1 = \bar{I}_2 + \bar{I}_m = 33.72\ \angle -4.59° + 15.02\angle -90° A$$

Hence,

$$I_1 = 38A$$

Slip speed in rpm at the rated torque and frequency:

$$N_{ss} = sN_s = 0.05\times 1500 = 75\ rpm$$

Since speed-torque curve is a straight line, slip speed at half the torque value:

Motor speed $N = 0.5 \times 75 = 37.5$ rpm

At 25 Hz:

$$N = \frac{25}{50}\times 1500 = 750\ rpm.$$

Since the slip remains constant for a given torque:

Motor speed, $N = N_s - N_{ss2} = 750-37.5 = 712.5$ rpm

Now for a constant flux the (v/f) ratio must be constant:

$$E = 0.5\times 480.55 = 240.28\ V \quad \therefore S = \frac{N_{ss2}}{N_s} = \frac{37.5}{750} = 0.05$$

$$Z_2 = \frac{R_2}{s} + jk\ X_2 = 7.6 + j0.57 = 7.62\ \angle 4.29°\ \Omega$$

Now, taking E as reference vector:

$$\bar{I}_2 = \frac{E}{Z_2} = \frac{240.28}{7.62\angle 4.29} = 31.53\angle -4.29°\,A.$$

$$\bar{I}_m = \frac{E}{jKX_m} = \frac{240.28}{j16} = 15.02\angle -90°\,A.$$

$$\bar{I}_1 = \bar{I}_2 + \bar{I}_{\bar{m}} = 31.53\angle -4.29° + 15.02\angle -90° = 35.92\angle -28.93°.$$

$$\bar{I}_1 = 35.92\,A$$

Now at the rated barking torque, the slip speed will be the negative of the slip speed at the rated motoring torque.

Therefore, the slip speed, $N_{ss3} = -75rpm$

Synchronous speed $\neq N + N_{ss3} = 1200 - 75 = 1125$ rpm

$$\text{Frequency} = \left(\frac{1125}{1500}\right) \times 50 = 37.5\,Hz. \quad \therefore K = \frac{37.5}{50} = 0.75.$$

At 37.5 Hz:

$$E = \left(\frac{37.5}{50}\right) \times 480.55 = 360.41\,V, S = \frac{-75}{1125} = -0.067$$

$$\bar{Z}_2 = \frac{R_2}{s} + jkX_2 = -5.67 + j\,0.855 = 5.73\angle 171.42°\Omega$$

Now taking E as reference vector:

$$\bar{I}_2 = \frac{E}{Z_2} = \frac{360.41}{5.73\angle 171.42°} = 62.9\angle -171.42°\ A$$

$\bar{I}_m$ remains the same as the forgoing:

$$\bar{I}_1 = 62.9\ \llcorner -171.42° + 15.02\ \llcorner -90° = -62.2 - j24.4 = 66.81\ \llcorner -158.58°.$$

$$\bar{V} = \bar{E} + \bar{Z}_1 \cdot \bar{I}_1 = 360.41 + (0.66\ + j0.75\ \times\ 1.14) \times\ 66.81\ \llcorner -158.58°$$

$$\bar{V} =\ 340.22 - j69.28 = 347.2\ \llcorner -11.51°$$

3.6 Output Performance Analysis of Type A Chopper and Four Quadrant Chopper Operation

3.6.1 Type A Chopper

Type A Chopper or First Quadrant Chopper

Type A chopper is shown in the below figure. It is also known as the first quadrant chopper. When the chopper is ON, Vo = Vs and the current flows in the direction of the load, but when the chopper is OFF, Vo is zero, but Io continues to flow in the same direction through the freewheeling diode FD. Thus, the average value of voltage and current, say Vo and Io will always be positive as shown in the below graph:

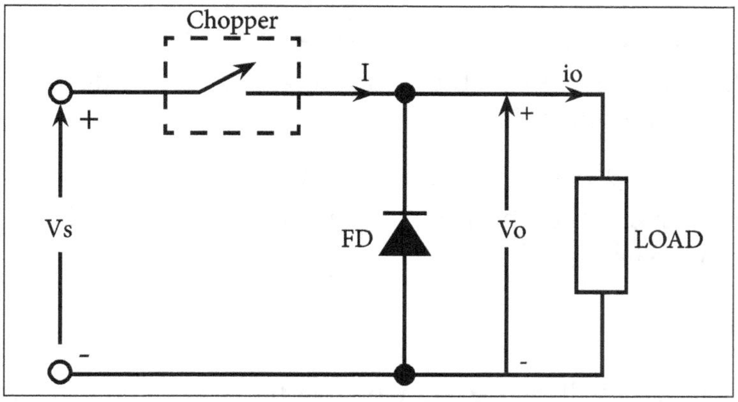

First quadrant chopper.

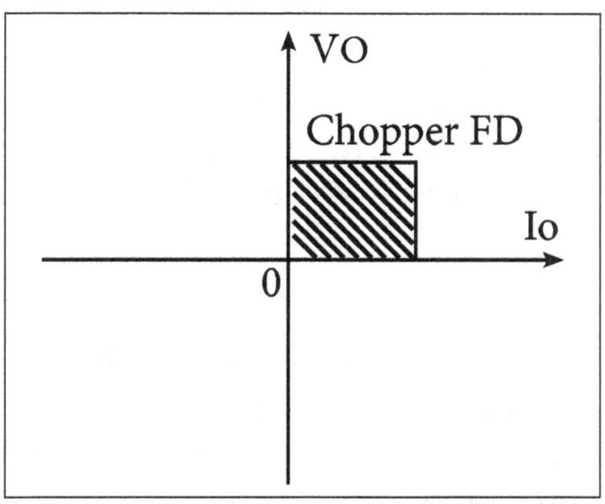

Graph of first quadrant chopper.

In type A chopper, the power will always flow from the source to load, as the average voltage Vo is less than the DC input voltage Vs.

3.6.2 Four Quadrant Chopper Operation

Type E Chopper or Fourth Quadrant Chopper

Type E or the fourth quadrant chopper consists of four semiconductor switches and four diodes arranged in anti-parallel. The four choppers are numbered according to the quadrant they belong. Their operation will be in each quadrant and the corresponding chopper only be active in its quadrant.

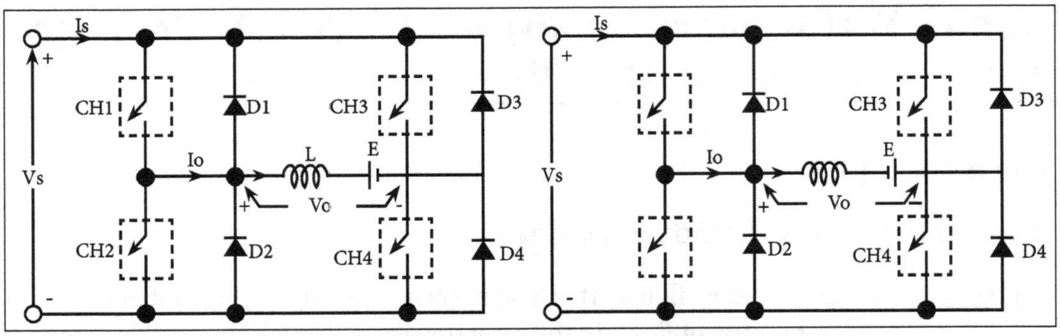

E type chopper circuit diagram with load emf E and E reversed.

First Quadrant

During the first quadrant operation, the chopper CH4 will be ON, CH3 will be OFF and chopper CH1 is operated. As CH1 and CH4 is ON, the load voltage Vo will be equal to the source voltage Vs and the load current io will begin to flow.

io and Vo will be positive, when first quadrant operation takes place. As soon as the chopper CH1 is turned OFF, positive current free wheels through the diode D2 and CH4. Type E chopper acts as a step down chopper in the first quadrant.

Second Quadrant

In this case, the chopper CH2 will be operational and other three choppers are kept OFF. As CH2 is ON, negative current will start flowing through CH2, E, and the inductor L and D4. Energy is stored in the inductor L, when the chopper CH2 is ON.

When CH2 is OFF, the current will be fed back to the source through the diodes D1 and D4. Here (E + L.di/dt) will be more than the source voltage Vs. In second quadrant, the chopper will act as a step up chopper, as the power is fed back from the load to source.

Third Quadrant

In third quadrant operation, CH1 will be kept OFF, CH2 will be ON and CH3 is operated. For this quadrant, the polarity of the load should be reversed. As the chopper CH3 is ON, the load gets connected to the source Vs and Vo and io will be negative and the third quadrant operation will take place. This chopper acts as a step down chopper.

Fourth Quadrant

In this quadrant, CH4 will be operated and CH1, CH2 and CH3 will be OFF. When the chopper CH4 is turned ON, positive current starts to flow through CH4, D2 and E and the inductor L will store the energy.

As CH4 is turned OFF, the current is fed back to the source through the diodes D2 and D3. The operation will be in fourth quadrant if the load voltage is negative, but the load current is positive. The chopper acts as a step up chopper as the power is fed back from the load to source.

4

Introduction to Motor Drives and Induction Motor Drives

4.1 Components of Power Electronic Drives

Motor drives are used in a very wide power range, from a few watts to many thousands of kilowatts, in applications ranging from very precise, high-performance position-controlled drives in robotics to variable-speed drives for adjusting the flow rates in pumps. In all the drives where the speed and position are controlled, a power electronic converter is needed as an interface between the input power and the motor.

Above a few hundred watt power level, there are basically the following three types of motor drives:

- DC motor drives.
- Induction motor drives.
- Synchronous motor drives.

A general block diagram for the control of motor drives is shown in the below figure. The process determines the requirements on the motor drive. For example, a servo-quality drive is needed in robotics, whereas only an adjustable-speed drive may be required in an air conditioning system.

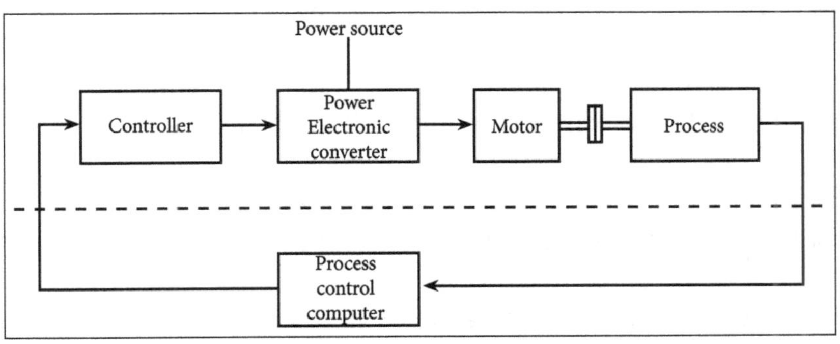

Control of motor drives.

In the servo applications of motor drives, the response time and the accuracy with which the motor follows the speed and position commands are extremely important. These servo systems, using one of these motor drives, require speed or position feedback for a precise control, as shown in the below figure.

In addition, if an AC motor drive is used, the controller must incorporate sophistication, such as field-oriented control, to make the AC motor meet the servo drive requirements. However, in a large number of applications, the accuracy and the response time of the motor to follow the speed command is not critical. As shown in the control of motor drives, there is a feedback loop to control the process, outside of the motor drive.

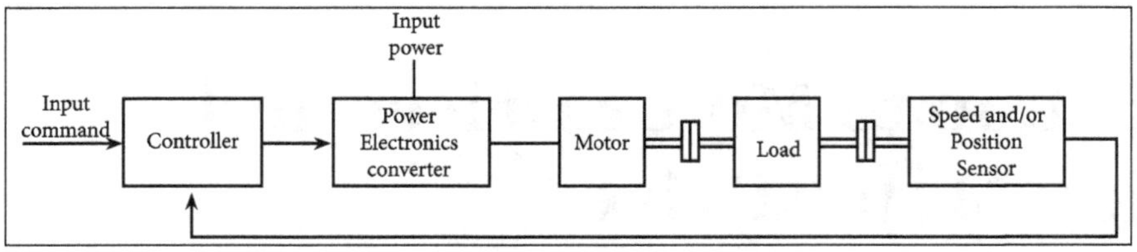

Servo drives.

Because of the large time constants associated with the process-control feedback loop, the motor drive's accuracy and the time of response to speed commands are not critical. An example of such an adjustable-speed drive is shown in the below figure for an air conditioning system.

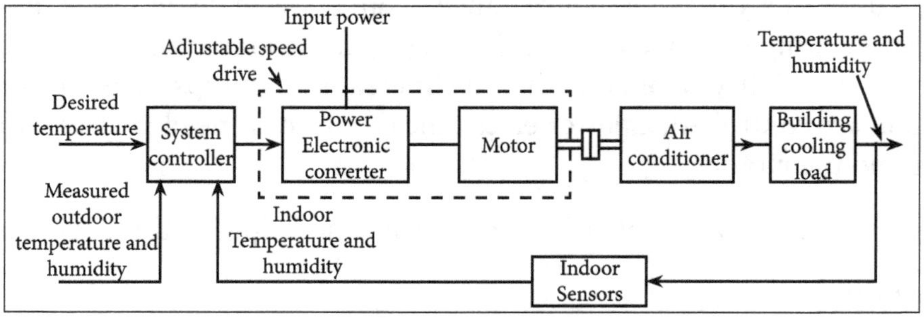

Adjustable speed drive in an air conditioning system.

4.2 Criteria for Selection of Drive Components: Match between the Motor and the Load

Factors Affecting the Selection of Electric Drive

The selection of electric drive basically means the selection of drive motor. Following are the various factors which influence the selection of motor to drive the load.

The limits of speed range: The range over which the speed control is necessary, for the load, similarly how hard is it to control the speed and the speed regulation also affects the choice of the motor.

The efficiency: The motor efficiency varies as load varies so the efficiency consideration under variable speed operation affects the choice of the motor.

The braking: The braking requirements from the load point of view, easy and effective braking are the requirements of a good drive.

Starting requirements: The starting torque necessary for the load, the corresponding starting current drawn by the motor also affects the selection of drive.

Power factor: It is well known that running of motors with low power factor values is not at all economical. While the power factor varies with the load conditions in some motors. Hence the type of the load and the running power factor of the motor are the essential considerations while selecting a drive.

Load factor: There is variety of types of load conditions possible like continuous, intermittent, impact etc. Such load variation factor and duty cycle of the motor influences the selection of drive.

Availability of supply: The motors available are a.c. or d.c. But the availability of supply decides the type of motor to be selected for the drive.

Effects of supply variations: There is a possibility of frequency supply variations. The motor selected must be able to withstand such supply variations.

Economical aspects: The size and rating of motor decides its initial cost while the various losses, temperature rise decides it's running cost. These economical aspects must be considered while selecting a drive.

Reliability of operation: It is important to study the conditions of stable operation of an electric drive. This includes the investigation of reliability of operation of an electric drive.

Environmental effects: It is possible that the atmosphere where an electric drive is to be used may contain some chemical gases, fumes, humidity etc. Such a contaminated atmosphere or a humid atmosphere also affects the choice of motor for a drive.

Selection of Motor for Miscellaneous Applications

There are various applications in practice for which a.c. or d.c. drives are used. Let us tabulate the various applications, the special features required and the corresponding type of motor to be selected.

S. No.	Application	Requirement	Type of motor
1.	Centrifugal pumps	Moderate starting torque without speed control	Standard squirrel cage
2.	Reciprocating pumps	100 to 200 % rated torque without speed control	Slip ring motor
3.	Crushers, punch presses, power hammers	Heavy starting torque and high peak loads	High torque squirrel cage or slip ring induction motor
4.	Stirrer for liquids	Light starting duty	Squirrel cage or slip ring induction motor
5.	Large ships	Large power demand	Synchronous or slip ring induction motor.
6.	Pulp grinders, load conveyors, line shafts	On load starting, high starting torque	Slip ring induction motor or high torque squirrel cage motor
7.	Machine tools	Fast acceleration, frequent starts and stops, effective and fast braking, last response.	Squirrel cage motor

8.	Domestic appliances mixers, sewing machines, vacuum cleaner	Simple speed control	Universal motor
9.	Domestic refrigerator compressors, air conditioners	Good starting torque	Single phase capacitor start induction motor
10.	Reciprocating compressors	Starting with partial load	Slip ring induction motor

Criteria for Selecting Drive Components

A motor drive consists of an electric motor, a power electronic converter, and possibly a speed or position sensor.

Match between the Motor and the Load

Prior to selecting the drive components, the load parameters and requirements such as the load inertia, maximum speed, speed range, and direction of motion must be available.

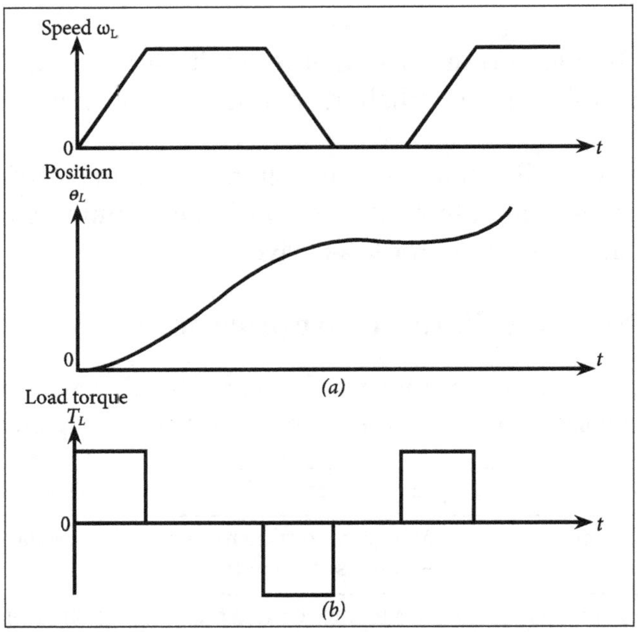

Load profile: (a) Load-motion profile; (b) Load-torque profile (assuming a purely inertial load).

The motion profile as a function of time must also be specified. By means of modeling the mechanical system, it is possible to obtain a load-torque profile. Assuming a primarily inertial load with a negligible damping, the torque profile, corresponding to the speed profile is shown in figure (a). The torque required by the load peaks during the acceleration and deceleration.

One way to drive the rotating load is to couple it directly to the motor. In such a direct coupling, the problems and the losses associated with a gearing mechanism are avoided. But the motor must be able to provide peak torques at specified speeds. The other option for a rotating load is to use a gearing mechanism. A coupling mechanism such as rack-and-pinion, belt-and-pulley, or feed-screw must be used to couple a load with a linear motion to a rotating motor. A gear and a feed-screw drive are shown in figure (a) and (b), respectively.

Assuming the energy efficiency of the gear in figure (a) to be 100%, the torques on the two sides of the gear are related as:

$$\frac{T_m}{T_L} = \frac{\omega_L}{\omega_m} = \frac{\theta_L}{\theta_m} = \frac{n_m}{n_L} = a \quad ...(1)$$

Where,

The angular speed $\omega = \dot{\theta}$

n_m and n_L are the number of teeth.

a is the coupling ratio.

In the feed-screw drive of figure (b), the torque and the force are related as:

$$\frac{T_m}{F_L} = \frac{v_L}{\omega_m} = \frac{x_L}{\theta_m} = \frac{s}{2\pi} = a \quad ...(2)$$

Where,

$v_L = \dot{x}_L$

The linear velocity

s is the pitch of the feed screw in m/turn.

u is the coupling ratio.

The electromagnetic torque T_{em} required from the motor can be calculated on the basis of energy considerations in terms of the inertias, required load acceleration, coupling ratio Q and the working torque or force.

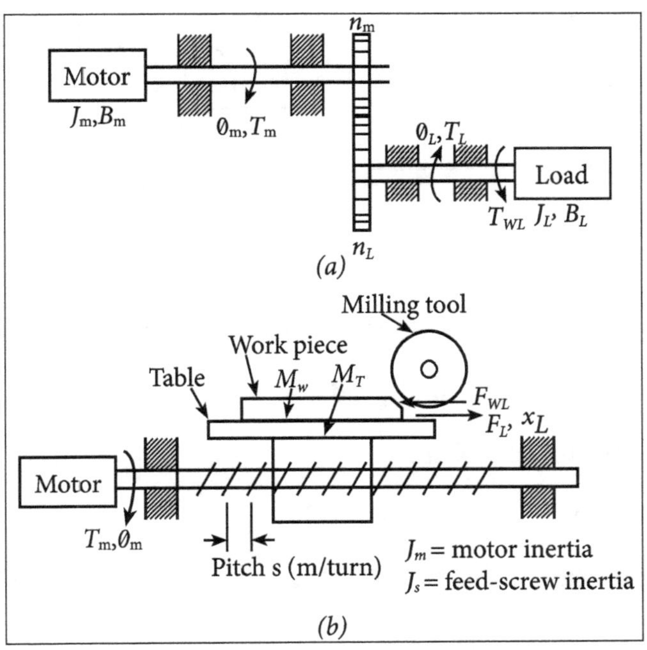

Coupling mechanisms: (a) Gear; (b) Feed.

In figure (a), T_{WL} is the working torque of the load and $\dot{\omega}_L$ is the load acceleration. Therefore:

$$T_{em} = \frac{\dot{\omega}_L}{a}\left[J_m + a^2 J_L\right] + aT_{wL} + \frac{\omega_L}{a}\left(B_m + a^2 B_L\right) \quad \text{...(3a)}$$

This equation can be written in terms of the motor speed, the equivalent total inertia $J_{eq} = J_m + a^2 J_L$, the equivalent total damping $B_{eq} = B_m + a^2 B_L$, and the equivalent working torque of the load $T_{Weq} = aT_{WL}$.

$$T_{em} = J_{eq}\dot{\omega}_m + B_{eq}\omega_m + Tw_{eq} \quad \text{...(3b)}$$

Similarly, for the feed-screw system in figure (b) with F_{WL} as the working or the machining force and a as the coupling ratio calculated in (2) in terms of pitch s, Tem can be calculated as:

$$T_{em} = \frac{\dot{v}_L}{a}\left[J_m + J_s + a^2\left(M_T + M_w\right)\right] + aF_{wL} \quad \text{...(4)}$$

Where $\dot{v}_L$ is the linear acceleration of the load.

As indicated by (1) and (2), the choice of the coupling ratio, a, affects the motor speed. At the same time, the value of affects the peak electromagnetic torque T_{em} required from the motor, as indicated by (3a) and (4). In selecting the optimum value of the coupling ratio a, the cost and losses associated with the coupling mechanism must also be included.

Fundamental Torque Equations of Motor-Load System

A motor generally drives a load (machine) through some transmission system. While the motor always rotates, the load may rotate or it may undergo a translational motion. Motor-load system by an equivalent rotational system is shown in the figure:

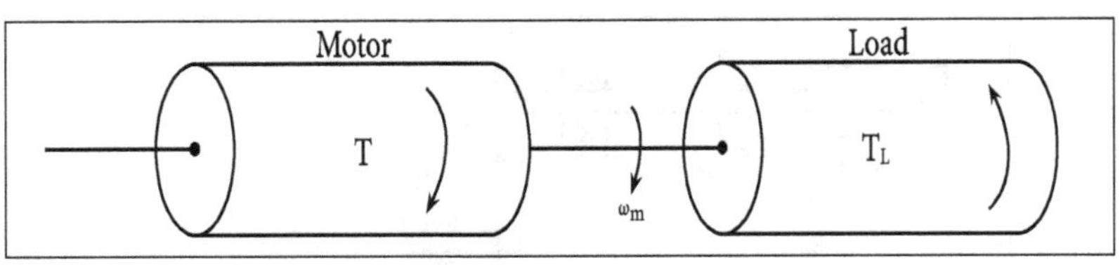

Motor load system.

Various Notations used are:

J = Polar moment of inertia of motor load system referred to the motor shaft, kg-m².

ω_m = Angular velocity of motor shaft, rad/sec.

T = Developed motor torque, N-m.

T_L = Load torque, referred to motor shaft, N-m.

Motor-load system of figure can be described by the following fundamental torque equation.

$$T - T_L = \frac{d}{dt}(J\omega_m) \quad \text{...(1)}$$

$$= J\frac{d\omega_m}{dt} + \omega_m \frac{dJ}{dt}$$

Equation (1) is applicable to variable inertia drives such as mine winders, reel drivers, industrial robots. For drives with constant inertia, $\left(\frac{dJ}{dt}\right) = 0$.

Therefore, $T - T_L = \frac{Jd\omega_m}{dt}$

$$T = T_L + \frac{Jd\omega_m}{dt} \qquad ...(2)$$

Equation (2) shows that torque developed by motor is counter balanced by a load torque 'T_L' and a dynamic torque $J\left(\frac{d\omega_m}{dt}\right)$. Torque component $J\left(\frac{d\omega_m}{dt}\right)$ is called the dynamic torque because it is present only during the transient operations.

Loads with Rotational Motion

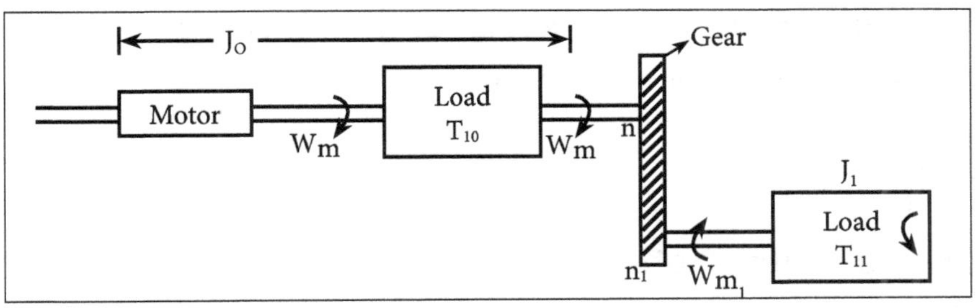

Motor load system.

Consider a motor driving two loads, one coupled direct to its shaft and other through a gear with n and n_1 teeth. The gear tooth ratio (a_1) is:

$$a_1 = \frac{n}{n_1} = \frac{w_{m1}}{w_m} = \frac{\text{Load speed}}{\text{Motor Speed}} \qquad ...(1)$$

If the losses in transmission are neglected, then the kinetic energy due to equivalent inertia must be same as kinetic energy of various moving parts.

$$\frac{1}{2}w_m^2 J = \frac{1}{2}J_o w_m^2 + \frac{1}{2}J_1 w_{m1}^2 \qquad ...(2)$$

Where,

J_o - Moment of inertia of motor and load directly coupled to the shaft (kgm²).

J_1 - Moment of inertia of load coupled through a gear (kgm²).

J - Moment of inertia referred to motor shaft (kgm²).

From equation (1) and (2):

$$J = J_o + a_1^2 J_1 \quad ...(3)$$

The total equivalent torque referred to motor shaft is (T_1).

$$T_1 w_m = T_{lo} w_m + \frac{T_{l1} wm_1}{\eta_1} \quad ...(4)$$

Where, η_1 = Transmission efficiency of the gears.

$$T_1 = T_{lo} + \frac{a_1 T_{l1}}{\eta_i} \quad ...(5)$$

[∵ from equation (1)].

If there are m other loads with moment of inertia's $J_1, J_2, ..., J_m$ and gear teeth ratios of $a_1, a_2, ..., a_m$ then:

$$J = J_o + a_1^2 . J_1 + ... + a_m^2 J_m .$$

If there are m loads with torques $T_{l1}, ..., T_{lm}$ and are coupled through gears with teeth ratios $a_1, a_2, ..., a_m$ and transmission efficiencies $\eta_1, \eta_2, ..., \eta_m$:

$$T_l = T_{10} + \frac{a_1 T_{l1}}{\eta_1} + ... + \frac{a_m T_{lm}}{\eta_m} \quad ...(6)$$

Loads with Translational Motion

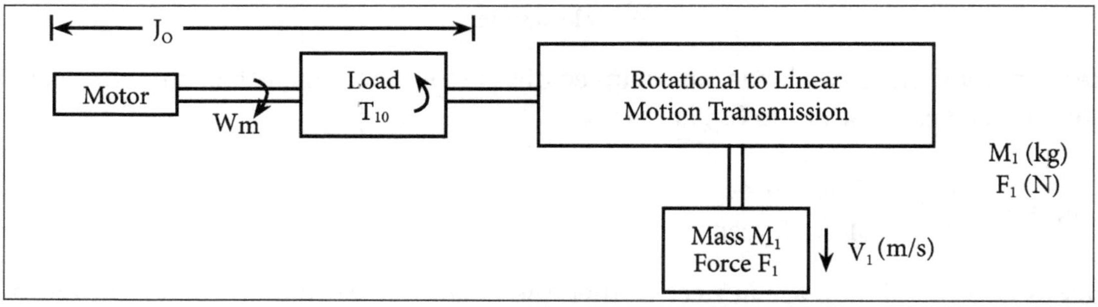

Consider a motor driving two loads, one coupled directly to its shaft and other through a transmission system.

If the transmission losses are neglected:

Kinetic energy due to J = Kinetic energy of various moving parts:

$$\frac{1}{2} J w_m^2 = \frac{1}{2} J_o w_m^2 + \frac{1}{2} M_1 V_1^2$$

$$J = J_o + M_1\left(\frac{V_1}{W_m}\right)^2$$

Load torque, $T_l = T_{lo} + \frac{f_1}{\eta_1}\left(\frac{V_1}{W_m}\right)$

If there are m loads with translational motion with velocities $V_1, V_2, ..., V_m$ and masses $M_1, M_2, ..., M_m$, then:

$$J = J_o + M_1\left(\frac{V_1}{W_m}\right)^2 + \ldots + M_m\left(\frac{V_m}{W_m}\right)^2$$

$$T_l = T_{lo} + \frac{f_1}{\eta_1}\left(\frac{V_1}{W_m}\right) + \ldots + \frac{f_m}{\eta_m}\left(\frac{V_m}{W_m}\right).$$

4.2.1 Thermal Consideration

The electromagnetic torque required from the motor as a function of time is obtained as shown in the below figure (a). In electric machines, the electromagnetic torque produced by the motor is proportional to the motor current i, provided the flux in the air gap of the motor is kept constant. Therefore, the motor-current profile is identical to the motor-torque profile, as shown in figure (b). The motor current in figure (b) during various time intervals is a DC current for a DC motor.

For an AC motor, the motor current shown is approximately the rms value of the AC current drawn during various time intervals. The power loss P_R in the winding resistance R_M, due to the motor current is the large part of the total motor losses, which gets converted into heat. This resistive loss is proportional to the square of the motor current and, hence, proportional to T^2_{em} during various time intervals in figure (a) and (b).

If the time period t_{period} in the below figure is short compared with the motor thermal time constant, then the motor heating and the maximum temperature rise can be calculated based on the resistive power loss P_R averaged over the time period t_{period}. Therefore, in the below figure, the rms value of the current over the period of repetition can be obtained as:

$$P_R = R_M I^2_{rms} \quad \text{...(1)}$$

Where,

$$I^2_{rms} = \frac{\sum_{k=1}^{m} I_k^2 t_k}{t_{period}} \quad \text{...(2)}$$

For example, consider m = 6.

Because of the motor current being linearly proportional to the motor torque, the rms value of the motor torque over t_{period} is given by:

$$T^2_{em,ems} = k_1 \frac{\sum_{k=1}^{m} I_k^2 t_k}{t_{period}} \quad ...(3)$$

Therefore:

$$T^2_{em}, rms = k_1 I^2_{rms} \quad ...(4)$$

Where k_1 is a constant of proportionality.

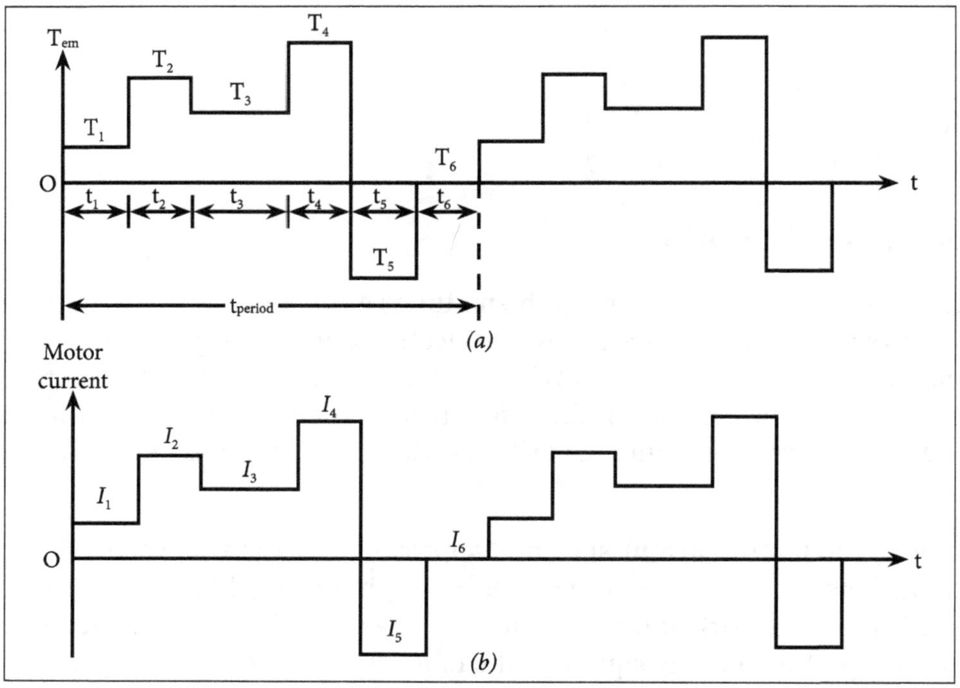

Motor torque and current.

From (1) and (4), the average resistive power loss P_R is given by:

$$P_R = k_2 T^2_{em,rms} \quad ...(5)$$

Where k_2 is a constant of proportionality.

In addition to PR, there are other losses within the motor that contribute to its heating. They are:

- P_s due to switching frequency ripples in the motor current.
- P_{EH} due to eddy currents and hysteresis within the motor laminations.
- P_{FW} due to friction and wind age.

Since it is supplied by a switching power electronic converter rather than an ideal source. There are always some powers losses termed as stray power losses P_{stray} that are not included with the foregoing losses. Therefore, the total power loss within the motor is given by:

$$P_{loss} = P_R + P_{FW} + P_{EH} + P_s + P_{stray} \quad ...(6)$$

Under a steady-state condition, the motor temperature rise $\Delta\Theta$ in degrees centigrade is given by,

$$\Delta\Theta = P_{loss} R_{TH} \quad ...(7)$$

Where,

P_{loss} is in watts.

The thermal resistance R_{TH} of the motor is in degrees centigrade per watt.

For a maximum allowable temperature rise $\Delta\Theta$, the maximum permissible value of P_{loss} in steady state depends on the thermal resistance R_{TH}. In general, the loss components other than P_R increase with the motor speed. Therefore, the maximum allowable P_R and hence, the maximum continuous motor torque output from (5) would decrease at higher speed, if R_{TH} remains constant.

However, in the self-cooled motors with the fan connected to the motor shaft, for example, R_{TH} decreases at higher speeds due to increased air circulation at higher motor speeds. Therefore, the maximum safe operating area in terms of the maximum rms torque available from a motor at various speeds depends on the motor design and is specified in the motor data sheets. For a motor torque profile, the motor should be chosen such that the rms value of the torque required from the motor remains within the motor's safe operating area in the speed range of operation.

4.2.2 Match between the Motor and the Power Electronics Converter

Depending on the power rating, speed of operation, operating environment, reliability, various other performance requirements by the load, and the cost of the overall drive, one of the following three types of motor drive such as DC motor drive, induction motor drive and synchronous motor drive is selected.

The power electronic converter topology and its control depend on the type of motor drive selected. In general, the power electronic converter provides a controlled voltage to the motor in order to control the motor current and hence, the electromagnetic torque produced by the motor.

Current Rating

The rms value of the torque that a motor can supply depends on its thermal characteristics. However, a motor can supply substantially larger peak torques provided that the duration of the peak torque is small when compared with the thermal time constant of the motor.

Since T_{em} is proportional to i, a peak torque requires a corresponding peak current from the power electronic converter. The current capability of the power semiconductor devices used in the converter is limited by the maximum junction temperature within the devices and other considerations.

A higher current results in a higher junction temperature due to power losses within the power semiconductor device. The thermal time constants associated with the power semiconductor devices are in general much smaller than the thermal time constants of various motors. Therefore, the current rating of the power electronic converter must be selected based on both the rms and the peak values of the torque that the motor is required to supply.

Voltage Rating

In both DC and AC motors, the motor produces a counter-emf e that opposes the voltage v applied to it, as shown by a simplified generic circuit. The rate at which the motor current and hence, the torque can be controlled is given by,

$$\frac{di}{dt} = \frac{v - e}{L}$$

Where L is the inductance presented by the motor to the converter.

To be able to quickly control the motor current and hence, its torque, the output voltage v of the power electronic converter must be reasonably greater than the counter-emf e. The magnitude of e in a motor increases linearly with the motor speed, with a constant flux in the air gap of the motor. Therefore, the voltage rating of the power electronic converter depends on the maximum motor speed with a constant air-gap flux.

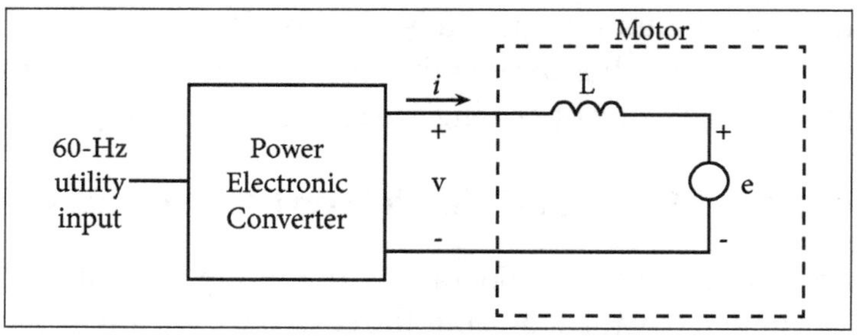

Simplified circuit of motor drive.

Switching Frequency and the Motor Inductance

In a servo drive, the motor current should be able to respond quickly to the load demand, thus requiring L to be small. Also, the steady-state ripple in the motor current should be as small as possible to minimize the motor loss P and the ripple in the motor torque. A small current ripple requires the motor inductance L to be large.

Because of the conflicting requirements on the value of L, the ripple in the motor current can be reduced by increasing the converter switching frequency. However, the switching losses in the power electronic converter increase linearly with the switching frequency. Therefore, a reasonable compromise must be made in selecting the motor inductance L and the switching frequency.

Selection of Speed and Position Sensors

In selecting the speed and position sensors, the following items must be considered:

- Direct or indirect coupling.

- Sensor inertia.
- Possibility and avoidance of torsional resonance.
- Maximum sensor speed.

To control the instantaneous speed within a specified range, the ripple in the speed sensor should be small. If such a sensor is used at very low speeds, the number of pulse outputs per revolution must be large to provide instantaneous speed measurement with sufficient accuracy. Similarly, accurate position information will require an incremental position encoder with a large number of pulse outputs per revolution.

4.2.3 Characteristics of Mechanical Systems: Stability Criteria

Equilibrium speed of a motor-load system can be obtained when motor torque equals the load torque. The drive system operates at steady state at this speed.

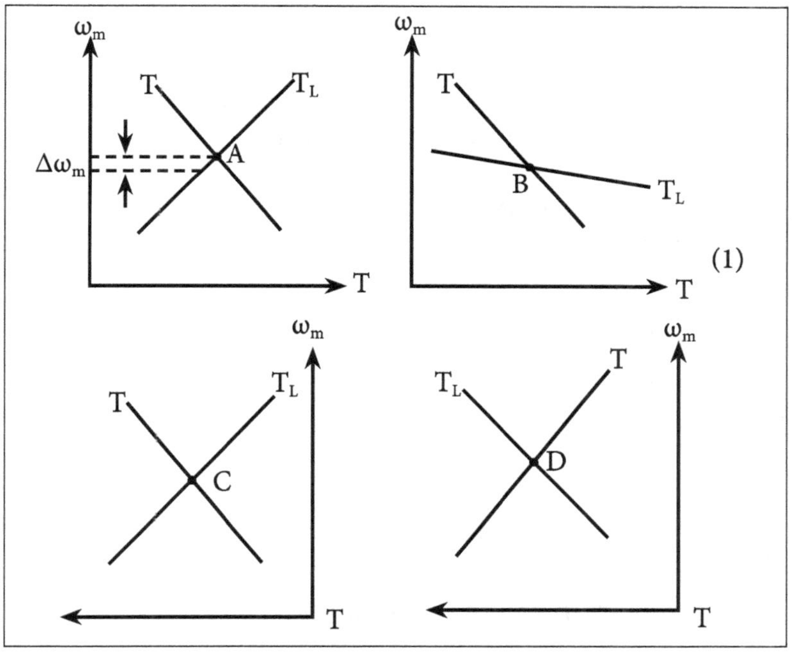

Torque - Speed characteristics.

- Point A and C are stable.
- Point B and D are unstable.

The equilibrium point will be termed as stable state when the operation will be restored to it after small departure from it due to disturbance in the motor or load. An increase of Δωm in speed caused by a disturbance will make load torque greater than the motor torque, resulting in deceleration and restoration of operation to point A and point C. Hence the steady state stable at point A and point C.

A decrease in speed causes the load torque to become greater than the motor torque, electric drive decelerate and operating point move away from point B. Increasing in speed will make T >> TL, which will move the operating point away from point B. Thus, point B is an unstable point of equilibrium.

From the above characteristics, an equilibrium point will be stable when an increase in speed causes load-torque to exceed the motor torque (i.e.,) at the equilibrium point following condition is satisfied.

$$\frac{dT_L}{d\omega_m} > \frac{dT}{d\omega_m}$$

Mathematical Condition for Steady State Stability

An equilibrium point will be stable when an increase in speed causes load torque (T_l) to exceed the motor torque (T). At equilibrium point, the following condition is satisfied.

$$\frac{dT_l}{d\omega_m} > \frac{dT}{d\omega_m}$$

The above inequality can be derived by an alternative approach. Let a small perturbation in speed $\Delta\omega$ results in ΔT and ΔT_l perturbations in T and T_l respectively.

$$(T+\Delta T)=(T_l+\Delta T_j)+J\frac{d(\omega_m+\Delta\omega_m)}{dt}$$

$$T+\Delta T=T_l+\Delta T_l+J\frac{d\omega_m}{dt}+J\frac{d\Delta\omega_m}{dt} \quad ...(1)$$

Actual motor torque equation is given by:

$$T=T_l+J\frac{d\omega_m}{dt} \quad ...(A)$$

Subtracting (A) from (1) and rearranging the terms gives:

$$J\frac{d\Delta\omega_m}{dt}=\Delta T-\Delta T_l \quad ...(2)$$

For small perturbations, the speed torque curves of the motor and load can be assumed to be straight lines. Thus:

$$\Delta T=\left(\frac{dT}{d\omega_m}\right)\Delta\omega_m \quad ...(3)$$

$$\Delta T_l=\left(\frac{dT_l}{d\omega_m}\right)\Delta\omega_m \quad ...(4)$$

Where, $\left(\frac{dT}{d\omega_m}\right)$ and $\left(\frac{dT_l}{d\omega_m}\right)$ are respectively slopes of the steady state.

Steady-state speed-torque curves of motor and load at operating point under consideration. Substituting equations (3) and (4) in (2) gives:

$$J\frac{d\Delta\omega_m}{dt}+\left(\frac{dT_l}{d\omega_m}-\frac{dT}{d\omega_m}\right)\Delta\omega_m=0 \qquad ...(5)$$

This is a first order linear differential equation. If initial deviation in speed at t = 0 be $(\Delta\omega_m)_0$. Then the solution of equation (5) will be:

$$\Delta\omega_m=(\Delta\omega_m)_0\exp\left\{-\frac{1}{J}\left(\frac{dT_l}{d\omega_m}-\frac{dT}{d\omega_m}\right)t\right\} \qquad ...(6)$$

An operating point will be stable when Δωm approaches zero as 't' approaches infinity. For this to happen the exponent in equation (6) should be negative.

Problems

1. Based on the mathematical condition, let us examine the stability equilibrium points given in figure.

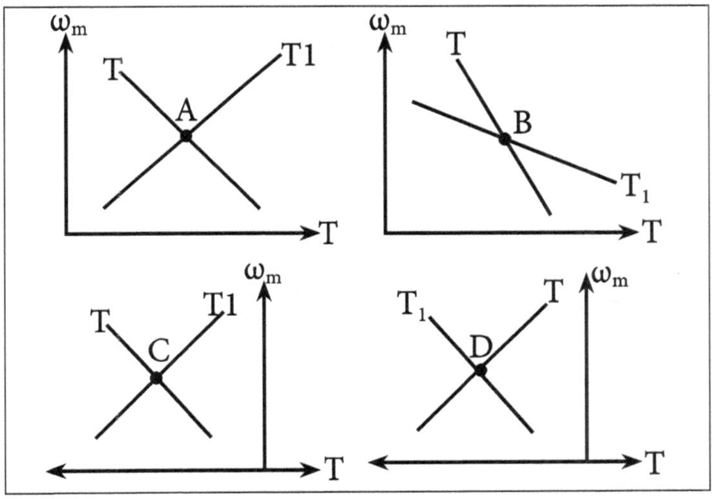

Solution:

Points	Examination of Stability
A	Stable
B	Unstable
C	Stable
D	Unstable

Examining steady state stability: The equilibrium point will be termed as stable when the operation will be restored to it after a small departure from it due to a disturbance in the motor or load.

Let the disturbance causes a reduction of $\Delta\omega_m$ in speed. At new speed, $T > T_1$, consequently motor will accelerate and operation will be restored to A. Similarly increase of $\Delta\omega_m$ will make $T_1 > T$, resulting in deceleration and restoration of operation to point A. Hence the drive is steady-state stable at point A.

At the equilibrium point B which is obtained when the same motor drives another load, a decrease in speed causes $T_1 > T$, drive decelerates and operating point moves away from B. Similarly, an increase in speed will make $T > T_1$, which will move the operating point away from B. Thus, B is an unstable point of equilibrium.

2. A drive has following parameters: J= 10 kg-m², T = 100 - 0.1 N-m. Passive load torque T_1 = 0.05 N, N-m where N is the speed in rpm. Initially the drive is operating in steady state. Now it is reversed. For this motor characteristic is T = -100 - 0.1N N-m. Let us calculate the time of reversal.

Solution:

Given:

$$J = 10\ \text{kg} - \text{m}^2$$

$$T = 100 - 0.1\ \text{N} - \text{m}$$

$$T_1 = 0.05\text{N}$$

Formula to be used

For Steady-State Speed, $T - T_1 = 0$

$$100 - 0.1\text{N} - 0.05\text{N} = 0$$

$$\text{N} = 666.1\ \text{rpm}$$

After reversal, for steady-state speed, noting that load is passive:

$$-100 - 0.1\text{N} - 0.05\text{N} = 0$$

$$\text{N} = -666.7\ \text{rpm}$$

$$J\frac{d\omega_m}{dt} = -100 - 0.1\text{N} - 0.05\text{N} \qquad ...(1)$$

Since, $\omega_m = \dfrac{2\pi N}{60} = \dfrac{\pi N}{30}$

$$(1) \Rightarrow \frac{dN}{dt} = \frac{30}{J\pi}(-100 - 0.15N) = -95.49 - 0.143N$$

$$\int dt = t = \int_{N1}^{N2} \frac{dN}{-95.49 - 0.143N} \qquad ...(2)$$

Where,

$$N_1 - 666.7\text{rpm}$$

$$N_2 = 0.95 \times -666.7$$

$$= -633.4 \text{ rpm.}$$

$$\Rightarrow t = 25.58 \text{ seconds.}$$

3. A motor is coupled to a load having the following characteristics.

$$\text{Motor}: T_m = 15 - 0.5\,\omega_m$$
$$\text{Load}: T_L = 0.5\,\omega_m{}^2$$

Let us find out the stable operating point for this condition.

Solution:

Given:

$$T_m = 15 - 0.5\,\omega_m,\ T_L = 0.5\,\omega_m^2 \qquad ...(1)$$

$$15 - 0.5\,\omega_m = 0.5\,\omega_m^2 \Rightarrow \omega_m = 5 \text{ and } -6.$$

Taking positive value for $\omega_m = 5$.

At $\omega_m = 5 \Rightarrow T_m = 12.5$ and $T_L = 12.5$.

ω_m(rad/sec)	2	4	5	6	7
T_m(N-m)	14	13	12.5	12	11.5
T_L(N-m)	2	8	12.5	18	24.5

The above table is obtained by substituting various ω_m values in equation (1).

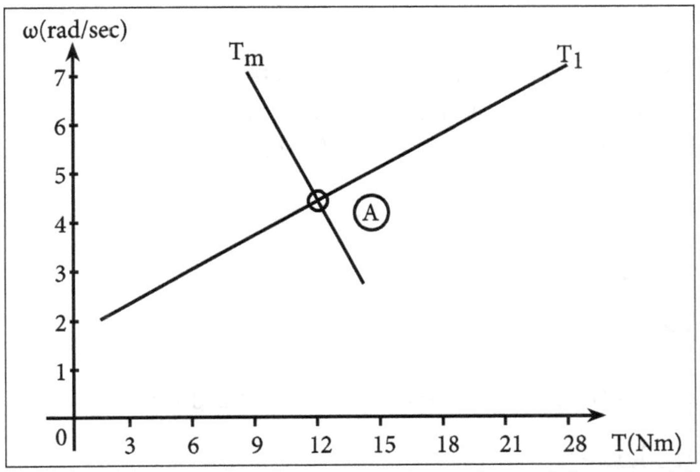

Point (A) is the equilibrium point. If we increase cu beyond the point (A) T_r Tm. If we decreases below point (A)Tj < T_m. So point (A) is a stable point.

4.3 Induction Motor Drives: Torque Speed Characteristics of 3-phase Induction Motor Drive

In this method of control, back-to-back thyristors are used to supply the motor with variable AC voltage. The analysis implies that the developed torque varies inversely as the square of the input RMS voltage to the motor. This makes such a drive suitable for fan and impeller-type loads for which torque demand rises faster with speed.

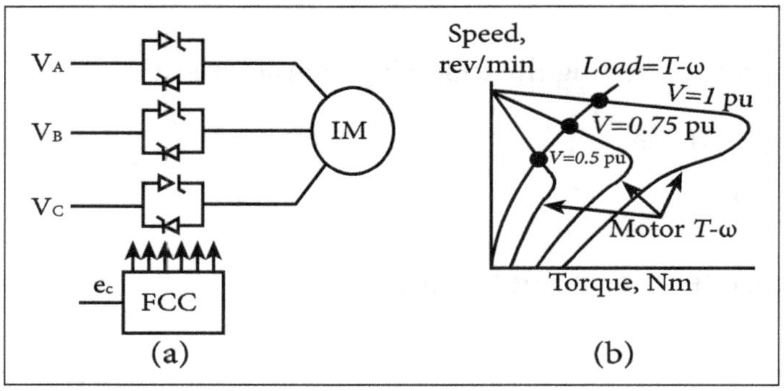

(a) Stator voltage controller. (b) Motor and load.

For other types of loads, the suitable speed range is very limited. Motors with high rotor resistance may offer an extended speed range. It should be noted that this type of drive with back-to-back thyristors with firing-angle control suffers from poor power and harmonic distortion factors when operated at low speed. If unbalanced operation is acceptable, the thyristors in one or two supply lines to the motor may be bypassed. This offers the possibility of dynamic braking or plugging, desirable in some applications.

Torque-speed Characteristics under Voltage Control

For the same slip and frequency, a small change in stator voltage results in a relatively large change in torque.

From the graph shown, the slip at the maximum torque remains unchanged since it is not a function of voltage.

For a low slip motor, the speed range is narrow. So this method is not used for wide range of speed control. It is excellent method for reducing starting current and increasing the efficiency during light load conditions. This method is only suitable for speed control below rated speed.

There are two different methods to achieve speed control using stator voltage control:

- Conventional method.
- Using auto transformer.
- Primary resistors connected in series with stator winding.
- AC voltage controller.

The induction motor speed variation can be easily achieved for a short range by either stator voltage control or rotor resistance control. But both of these schemes result in very low efficiencies at lower speeds.

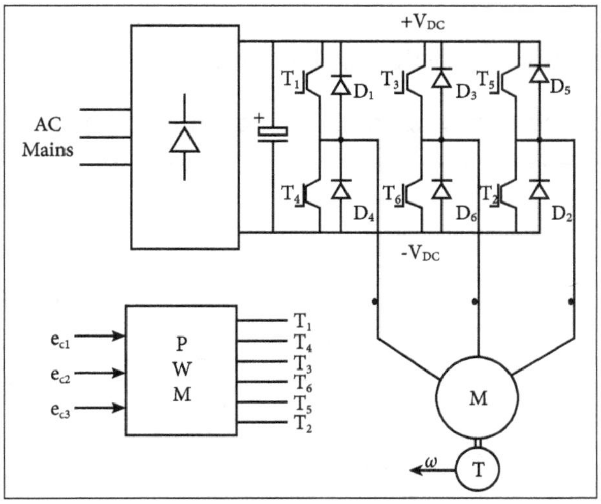

AC voltage controller.

The most efficient scheme for speed control of induction motor is by varying the supply frequency. This will not only results in scheme with wide speed range but also improves the starting performance. If the machine is operating at speed below base speed, then v/f ratio is to be kept constant so that the flux remains constant.

This retains the torque capability of the machine at the same value. But at lower frequencies, the torque capability decrease and this drop in torque has to be compensated for increasing the applied voltage.

4.3.1 Speed Control of 3-phase Induction Motor by Varying Stator Frequency and Voltage

v/f Control of Induction Motor Drives

The voltage/frequency control method is one of the methods of speed control of 3φ induction motor. The increase in the supply frequency increases the motor speed and also reduces the maximum torque. But increase in voltage results in an increase in maximum torque of the motor.

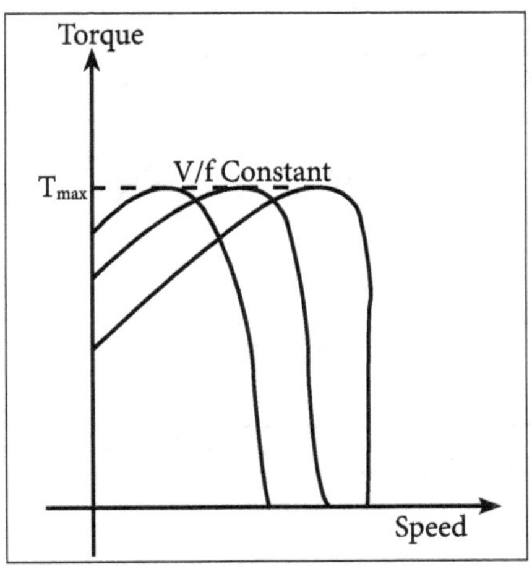

Torque vs Speed.

The maximum torque should be constant by keeping fixed v/f ratio, it is given by:

$$T_{max} \approx k(v/f)^2.$$

T_{max} is maintained constant and air gap flux is constant.

Open loop v/f control of induction motor can be obtained from the figure.

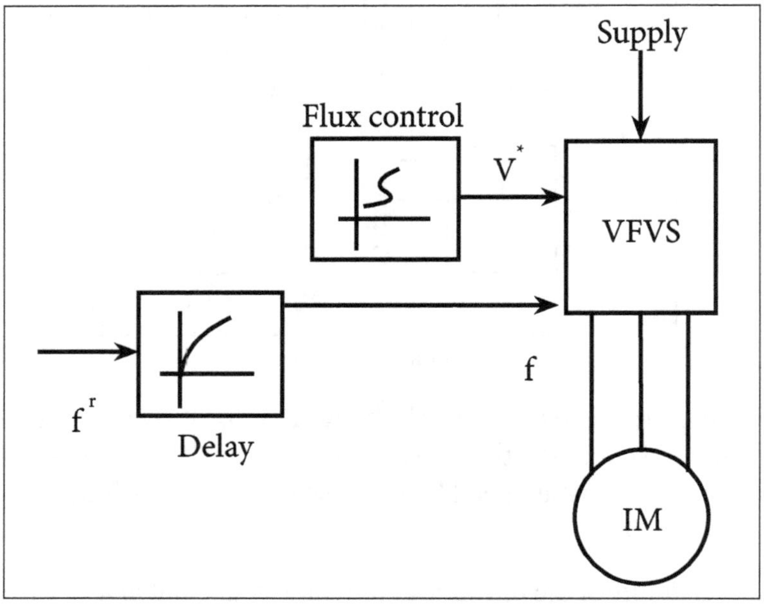

Open loop v/f control of induction motor.

The induction motor is fed from variable frequency variable voltage source (VFVS). The v* and f* are voltage and frequency commands for VFVS which is given by flux control and delay circuit respectively. Therefore, VFVS maintains the v/f ratio constant.

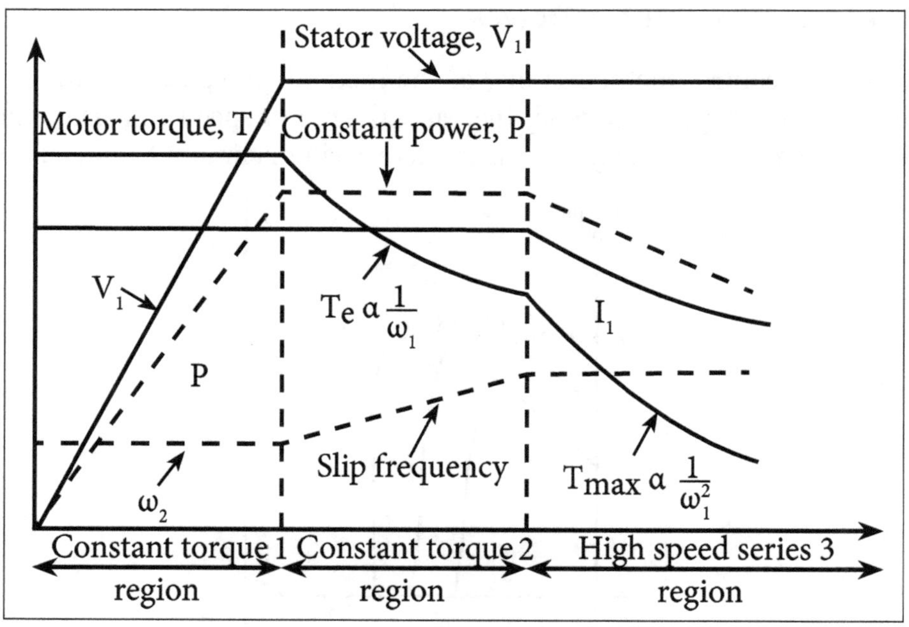

Stator Voltage, current, slip frequency, torque and power variation with speed-torque characteristics.

Theory of Voltage/Frequency Control

The voltage/frequency (v/f) control method is used for speed control of 3-phase induction motor. The relation between maximum torque and voltage/frequency is given by:

$$T_{max} = C\left(\frac{v}{f}\right)^2 \quad ...(1)$$

Where,

T_{max} = Maximum torque (N-m).

C = Constant.zx.

v = Voltage (volts).

f = Frequency (Hz).

From equation (1) we can say that:

- If frequency (f) increases, speed (N) increases and maximum torque (T_{max}) decreases. ($\therefore$ f α N).
- If voltage increases then maximum torque (T_{max}) also increases.

By combining the above 2 features, we can achieve a control design by which speed increases and torque is kept same. This is called as voltage/frequency control.

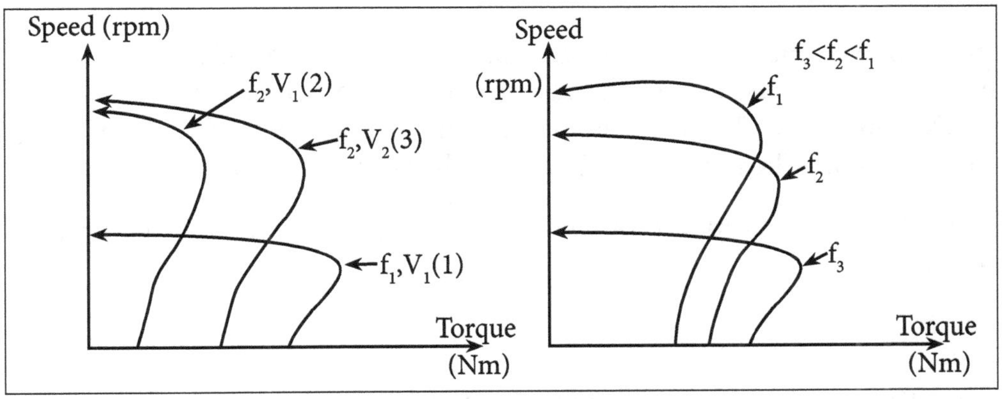

Speed-torque characteristics.

$V_1, F_1 \rightarrow$ reference voltage and frequency.

- If the supply frequency is increased to f_2(new) while keeping the voltage (V_1) unchanged, the motor speed increases and maximum torque decreases. The load is higher than T_{max} provided by motor. Thus no steady-state operating point can be achieved and motor eventually stalls. Curve (2) in the above figure describes this point.
- In this case, if the magnitude of the voltage is increased to V_2(new) then the motor characteristics stretches and T_{max} increases. A new steady-state point is achieved. Curves (3) in above figure describe this point.

The low frequency operation at a constant voltage/frequency ratio is not satisfactory. Due to this very low frequency, the no-load current increases and the starting and breakdown torque decreases.

Therefore for maintaining voltage/frequency constants, we can increase the supply voltage. The increased voltage can cause damage to insulation's of motors. So,

$$(\text{Voltage}) \leq 1.1\,(V_{rated})$$

V_{rated} → Rated voltage (volts).

The cyclo converters or the 3-phase inverters have the ability to provide the voltage at variable frequency.

4.3.2 Impact of Non-sinusoidal Excitation on Induction Motors

It was assumed that the induction motors are supplied from a three-phase, balanced, and sinusoidal set of voltages. In practice, the inverters used in variable-frequency controllers produce three phase voltages or currents that are identical in each phase, except for the 120° phase displacement. Unfortunately, these are not purely sinusoidal and contain higher frequency components that are harmonics of the fundamental frequency.

In the following analysis, we will assume that the motor is supplied by three-phase voltage sources, as in the case of a voltage source inverter. This analysis can be easily modified to three-phase current sources, as in the case of a current source inverter.

Harmonic Motor Currents

As a first-order approximation, the motor currents in the presence of harmonic voltage components can be found by calculating each harmonic current component i_h (at harmonic h) from the per-phase equivalent circuit. Then the motor currents can be obtained by using the principle of superposition and adding the fundamental and all other harmonic current components.

At the harmonic h, the flux produced by the voltage components (v_{ah}, v_{bh}, v_{ch}) rotates in the air gap at a speed of:

$$\omega_{sh} = h\omega_s$$

Where, the direction may be same or in opposition to the rotor's direction of rotation.

It can be easily verified that the flux produced by the harmonics h = 6n - 1 (where, n = 1, 2, 3, .. .) has an opposite phase rotation compared with the fundamental. Therefore, these harmonics result in a flux rotation opposite to that of the rotor.

Harmonics, h = 6n + 1 (where n = 1, 2, 3, . . .) produces a flux rotation in the same direction as the rotor. Under a variable-frequency operation for speed control, the motor rotates at a reasonably small value of slip. Thus, to the first-order approximation, the rotor speed can be assumed to be same as the fundamental-frequency synchronous speed,

$$\omega_r \approx \omega_s$$

Therefore, at a harmonic h in the equivalent circuit, the rotor slip relative to the synchronous speed at the harmonic frequency is given by,

Where, the direction may be same or in opposition to the rotor's direction of rotation.

It can be easily verified that the flux produced by the harmonics h = 6n - 1 (where, n = 1, 2, 3, .. .) has an opposite phase rotation compared with the fundamental. Therefore, these harmonics result in a flux rotation opposite to that of the rotor.

Harmonics, h = 6n + 1 (where n = 1, 2, 3, . . .) produces a flux rotation in the same direction as the rotor. Under a variable-frequency operation for speed control, the motor rotates at a reasonably small value of slip. Thus, to the first-order approximation, the rotor speed can be assumed to be same as the fundamental-frequency synchronous speed:

$$\omega_r \approx \omega_s$$

Therefore, at a harmonic h in the equivalent circuit, the rotor slip relative to the synchronous speed at the harmonic frequency is given by:

$$\text{Slip } s_h = \frac{\omega_{sh} \pm \omega_r}{\omega_{sh}} \simeq \frac{h \pm 1}{h} \approx 1$$

Where the ± sign corresponds to the direction of air gap flux rotation in opposition to, or same as the rotor's direction of rotation, respectively.

Recognizing that $\omega_r \approx \omega_s$ and $s_h = 1$, an approximate equivalent circuit at the frequency of harmonic h is shown in the per-phase harmonic equivalent circuit, which is obtained by neglecting L.

If the motor is excited from a voltage source and the harmonic components of the stator voltage v are known, the corresponding harmonic components in the motor current i, can be obtained by using the principle of superposition and the harmonic equivalent circuit for each harmonic, one at a time.

For calculating the harmonic current components, the magnetizing components can be neglected and the harmonic current magnitude is primarily determined by the leakage reactances at the harmonic frequency, which dominate over R_s and R_r.

$$I_h \approx \frac{V_h}{h\omega\left(L_{ls} + L_{lr}\right)}$$

The above equation for I_h shows that by increasing the frequencies at which the harmonic voltages occur in the converter output, the magnitudes of harmonic currents can be reduced. Note that the foregoing procedure for calculating the harmonic currents is the first-order approximation, since the motor leakage reactance and resistances may vary with the frequency.

Harmonic Losses

The additional losses which are present in the induction motor due to the harmonics in the supply voltage waveform are termed as harmonic losses. The stator copper loss increases not only due to the existence of harmonic currents, but also due to the increase in the fundamental current itself. The resistance of the stator winding also increases due to skin effect, especially for higher-order harmonic currents flowing in the stator windings of the medium-power and large-power motors.

The additional rotor copper loss is primarily due to the significant increase in the rotor resistance due to skin effect, particularly in the case of deep-bar rotors. With a constant-torque load, a very small increase in the fundamental rotor current may occur to overcome the small net negative torque produced by the harmonic currents. The core loss in the motor also increases slightly due to the higher peak value of the flux density wave caused by non-sinusoidal excitation.

The end-leakage and skew-leakage fluxes, which normally contribute to the stray load loss, may produce an appreciable core loss at the harmonic frequencies. The friction and windage loss remains the same, since it depends only on the speed of rotation and not on the voltage waveform.

Efficiency of the Motor

For a given load torque to be operated at a desired speed, the efficiency of the motor depends only on the total losses in the motor. The magnitude of the additional losses created by the harmonics depends essentially on the nature of the applied voltage waveform.

Most static inverters do not generate harmonics of lower order than the fifth and the higher-order ones have small amplitudes. For motors fed from such inverters, only a slight reduction in its efficiency occurs.

Even if there occurs an overall increase of 25% in motor losses, a motor having an efficiency of 88% on a 50 Hz sinusoidal supply will have its losses increased from 12% to 15%, thereby having only a 3 percent reduction in its efficiency.

4.4 Classifications of Variable Frequency Converter

A three-phase variable frequency inverter supplying an induction motor is shown in the figure. The power devices are assumed to be ideal switches. There are two major types of switching schemes for the inverters namely, square wave switching and PWM switching.

The converter which converts the power into AC power is commonly known as a inverter. The application areas for the inverter include the Uninterrupted Power Supply (UPS), AC motor speed controllers, etc.

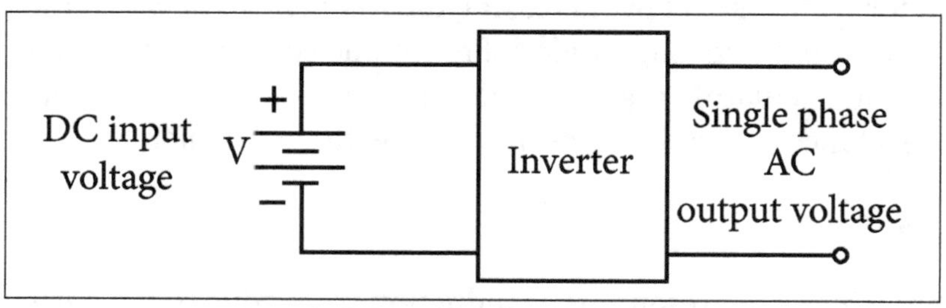

Block diagram of an inverter.

The inverters can be classified based on a number of factors like, the nature of output waveform, the power devices being used, the configuration being used, the type of commutation circuit that is being employed and the voltage source and current source inverters.

Thyristorized inverters use SCRs as power switches, because the input source of the power is pure DC in nature, forced commutation circuit is an essential part of thyristorized inverters. The commutation circuit must be carefully designed in order to ensure successful commutation of SCRs. The addition of commutation circuit makes the thyristorized inverters bulky and costly.

The size and cost of the circuit can be reduced to some extent if the operating frequency is increased, but the inverter grade thyristors, which are special thyristors, are manufactured to operate at a higher frequency, which are costly.

Inverters can be classified on the basis of a number of factors such as:

Classification According to the Nature of Input Source

Based on the nature of input power source, inverters are classified as:

- Voltage source inverters (VSI).
- Current source inverters (CSI).

In case of VSI, the input to the inverter is provided by a ripple free DC voltage source whereas in CSI, the voltage source is first converted into a current source and then used to supply the power to the inverter.

Classification According to the Wave Shape of the Output Voltage

The inverters can be classified according to the nature of output voltage waveform as:

- Square-wave inverter.
- Quasi-square wave inverter.
- Pulse-width modulated (PWM) inverters.

Types of Variable Frequency Drives

There are three common Variable Frequency Drives (VFDs) that offer both advantages and disadvantages depending on the application they are used for. The three common VFD designs used include:

- Current Source Inverter (CSI).
- Voltage Source Inverter (VSI).
- Pulse Width Modulation (PWM).

However, there is a fourth type of VFD called Flux Vector Drive, which is emerging in popularity among end-users for its closed-loop control feature. Each VFD consists of a Converter, DC Link and Inverter section but each one's construction varies from drive to drive. Although the sections of each VFD are similar, they require a variation in circuitry in supplying the frequency and voltage to the motor.

4.4.1 Variable Frequency PWM VSI Drives

PWM Principle

The simplest way to control the speed of the motor is to use pulse width modulation. The below figure shows the basic structure of a pulse modulated wave. By using pulse width modulation, the power loss occurred during the switching operation is reduced as the transistor is either fully ON or fully OFF. This results in much lesser power dissipation and a stabilized speed.

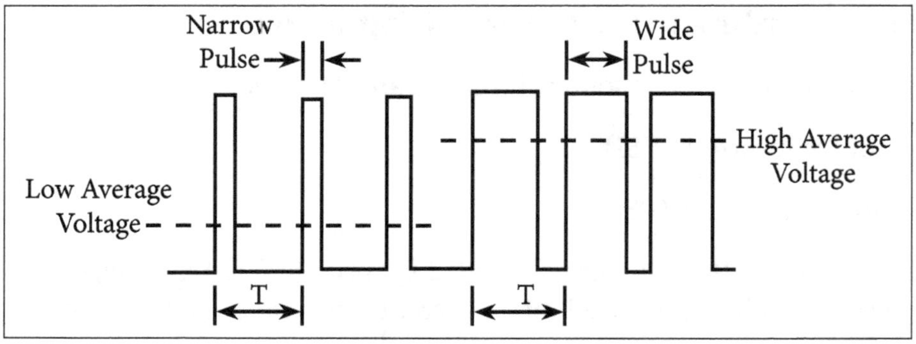

Pulse width modulated waveform.

By using a basic 555 oscillator circuit in astable mode, the PWM signal can be generated which can be used for controlling the speed of the motor.

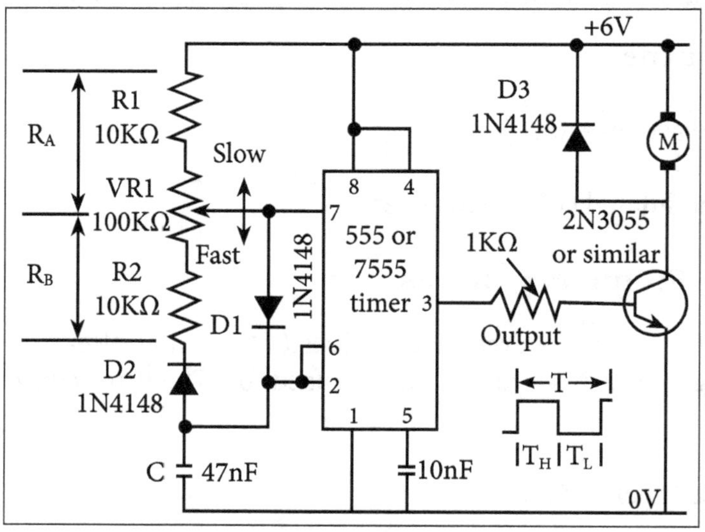

Oscillator circuit in astable mode.

This simple circuit based around the familiar NE555 or 7555 timer chip is used to produce the required pulse width modulation signal at a fixed frequency output. The timing capacitor C is charged and discharged by current flowing through the timing networks R_A and R_B.

The output signal at pin 3 of the 555 is equal to the supply voltage switching the transistors fully "ON". The time taken for C to charge or discharge depends upon the values of R_A, R_B.

The capacitor charges up through the network R_A but gets diverted around the resistive network R_B and through diode D1. When the capacitor is charged, it gets discharged through diode D2 and network R_B into pin 7. During the discharging process the output at pin 3 is at 0 V and the transistor is switched "OFF".

The time taken for capacitor C to finish one complete charge-discharge cycle depends on the values of R_A, R_B and C, the time T for one complete cycle is given as:

- The time, T_H, for which the output is "ON" is: $T_H = 0.693(R_A).C$.
- The time, T_L, for which the output is "OFF" is: $T_L = 0.693(R_B).C$.
- Total cycle time $T = T_H + T_L$.
- Output frequency $f = 1/T$.

The duty cycle of the waveform can be adjusted from about 8.3% (0.5V) to about 91.7% (5.5V) using a 6.0V power supply. The Astable frequency is constant at 256Hz and the motor is switched "ON" and "OFF" at this rate.

Resistor R_1 represent the top part of the potentiometer, VR_1 represent the resistive network of R_A. The bottom part of the potentiometer R_2 represent the resistive network of R_B above. These values can be changed to suite different applications of DC motors. There should be no jerkiness in the rotation of the motor.

Diode D3 is the flywheel diode used to protect the electronic circuit from the inductive loading of the motor. If the motor load is high a heat sink on the switching transistor or MOSFET can be added.

4.4.2 Variable Frequency Square Wave VSI Drives

Square Wave Inverters

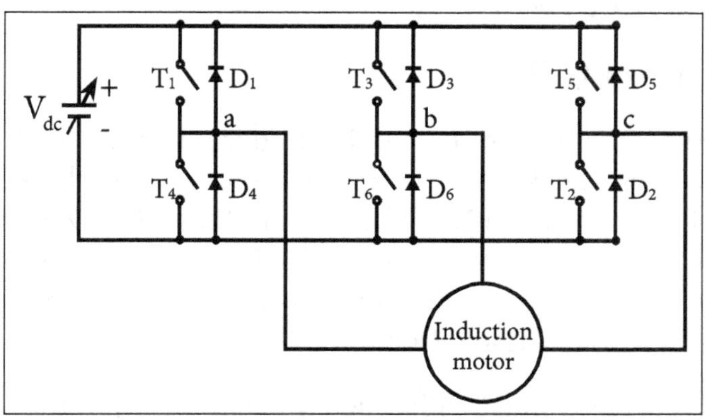

A schematic of the generic inverter-fed induction motor drive.

The gating signals and the resulting line voltages for square wave switching are shown in the figure. The phase voltages are derived from the line voltages assuming a balanced three-phase system.

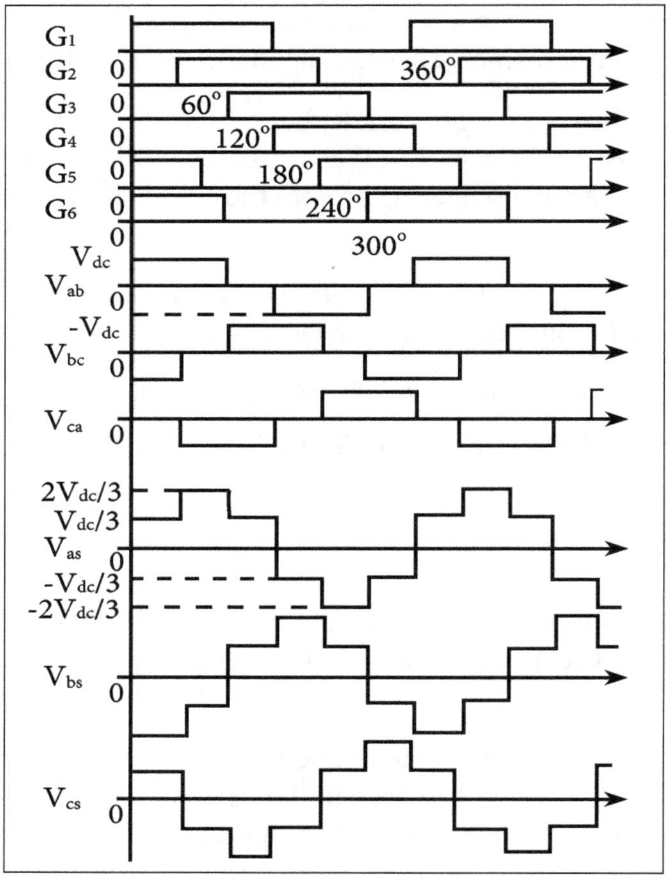

Inverter gate (base) signals and line-and phase-voltage waveforms.

The square wave inverter control is simple and the switching frequency and consequently, switching losses are low. However, significant energies of the lower order harmonics and large distortions in current wave require bulky low-pass filters. Moreover, this scheme can only achieve frequency control. For voltage control a controlled rectifier is needed, which offsets some of the cost advantages of the simple inverter.

4.4.3 Variable Frequency CSI Drives

Current Source Inverter (CSI)

A Current Source Inverter (CSI) is a type of variable frequency drive (VFD) which converts the incoming AC voltage and varies the frequency and voltage supplied to the AC Induction Motor. The general configuration of this type of VFD is like that of other VFDs which consists of a Converter, DC Link and Inverter. The converter part of the CSI uses silicon-controlled rectifiers (SCRs), gate-commutated thyristors (GCTs) or symmetrical gate-commutated thyristors (SGCTs) to convert the incoming AC voltage to a variable DC voltage.

In order to maintain the correct voltage to frequency (Volt/Hertz), the voltage must be regulated by the correct sequencing of the SCRs. The DC Link for this type of variable frequency drive uses an inductor to regulate the current ripple and to store the energy used by the motor.

The inverter, which may be responsible for converting the DC Voltage back to an AC sine-like waveform, comprises of SCRS, gate turn-off thyristors (GTOs) or symmetrical gate-commutated thyristors (SGCTs). These thyristors behave like switches which are turned on and off to create the pulse width modulation (PWM) output that regulates the frequency and voltage to the motor.

CSI variable frequency drives regulate current; require a large internal inductor and a motor load to operate. In CSI VFD designs, the requirements of input and output filters are necessary due to high harmonics in the power input and poor power factor. To resolve this issue, many manufacturers implement either input transformers or reactors and harmonic filters at the point of common coupling to reduce the harmonics on the drive system.

Of the common VFD drive systems, CSI VFDs are the only type of drives that have regenerative power capability. Regenerative power capability means that power is driven back from the motor to the power supply can be absorbed.

Advantages of CSI

- Simple circuitry.
- Regenerative power capability.
- Reliability (Current Limiting Operation).
- Clean current waveform.

Disadvantages of CSI

- Motor cogging when PWM output is below 6 Hz.
- Inductors used are large and costly.
- Large power harmonic generation sent back into power source.
- Low input power factor.

CSI Fed Induction Motor Drive

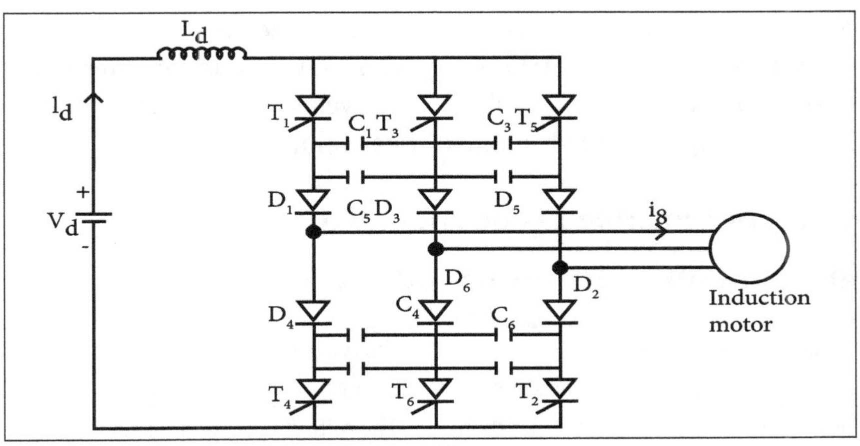

VSI drive

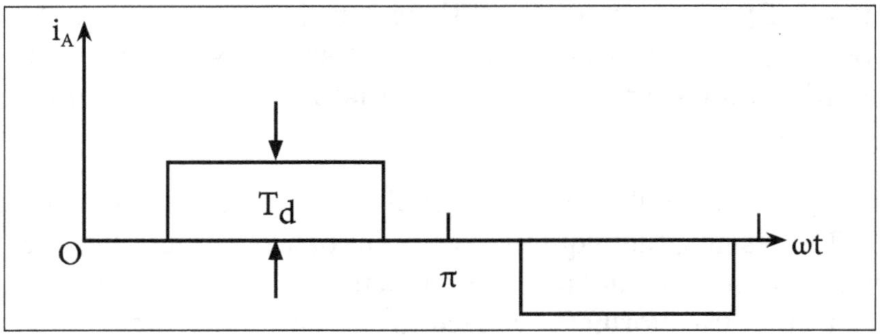

Waveform

$$I_S = \frac{\sqrt{6}}{\pi} I_d$$

Diodes D_1 – D_6 and capacitors C_1 – C_6 provide commutation of thyristors T_1 – T_6, which are fired with a phase difference of 60° in sequence of their numbers. It also shows the nature of output current wave forms.

For a given speed, torque is controlled by varying DC link current I_d by changing the value of V_d. Therefore, when supply is AC, a controlled rectifier is connected between the supply and inverter and when supply is DC, a chopper is interposed between the supply and inverter.

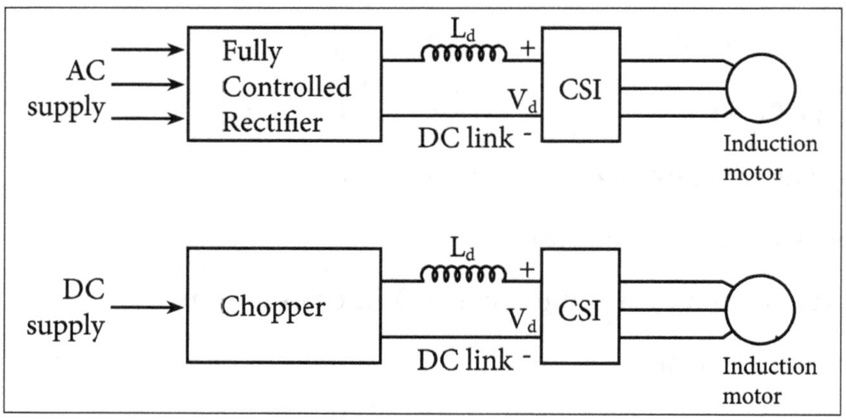

CSI fed induction motor drive.

In CSI, conduction of 2 devices in the same leg does not lead to sudden rise of current through them due to the presence of a large inductance L_d. This allows time for commutation to take place and normal operation to get restored in subsequent cycles. The commutation capacitors C_1 - C_6 reduces the voltage spikes by reducing the rate of rise and fall of current.

4.4.4 Comparison of Variable Frequency Drives

Difference between Voltage/Current Fed Inverters

S.No.	Voltage Fed Inverters	Current Source Inverters
1	The input voltage is maintained constant.	Input current is constant but-adjustable.
2	The output voltage does not depend on the load.	The output current does not depend on the load.

3	The magnitude of output current depends upon on load impedance.	The magnitude of output voltage depends upon the load impedance.
4	It requires feedback diodes.	It does not require any feedback diode.
5	Commutation circuit is complex.	Commutation circuit is simple.
6	The power factor is low and poor at low load.	The power factor is high.

4.4.5 Line Frequency Variable Voltage Drives

The line-frequency variable-voltage drive is used for induction motors which have large rotor resistance and hence, pull-out torque at large slip values. The drive permits speed control over a wide range. The electromagnetic torque in terms of the AC source voltage is given by:

$$T_{em} \propto f\phi_{ag}{}^{2} \propto \frac{V_S^2}{f}$$

For constant-torque operation, the speed is controlled by varying the source voltage. Speed control results in poor efficiency at low speeds because of dissipation in large rotor resistance. This technique is widely used in the fractional horsepower fan or pump drives, cranes and hoist drives which may have intermittent-load operation.

4.5 Soft Start of Induction Motors

Starting current in the induction motors may be approximately six times the rated current. The soft start of the induction motor, reduced current starting is actually the reduced voltage starting. The circuit in the below figure can also be used for soft starting in constant speed control. Each thyristor conducts for an entire half cycle. The back-to-back thyristor pairs are shorted out of the circuit by mechanical contractors in the steady-state condition.

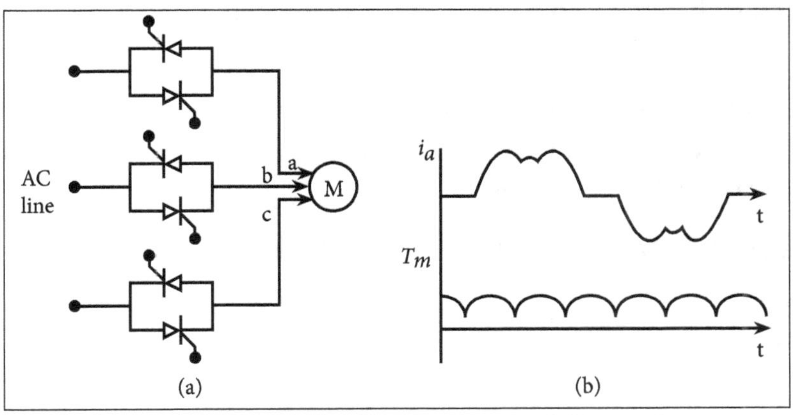

Soft start of induction motors: (a) Circuit and (b) waveforms.

Constant V/F Control of Induction Motor

This is the most popular method for controlling the speed of an induction motor. As in v/f method, if the supply frequency is reduced keeping the rated supply voltage, the air gap flux will tend to saturate. This will cause excessive stator current and distortion of the stator flux wave. Therefore, the stator voltage should also be reduced in proportional to the frequency so as to maintain the air-gap flux constant.

The magnitude of the stator flux is proportional to the ratio of the stator voltage and the frequency. Hence, if the ratio of voltage to frequency is kept constant, the flux remains constant. Also, by maintaining V/F as constant, the developed torque remains approximately constant.

This method gives higher run-time efficiency. Therefore, majority of AC speed drives employ constant V/F method (or variable voltage, variable frequency method) for the speed control. Along with wide range of speed control, this method also offers 'soft start' capability.

4.5.1 Speed Control by Static Slip Power Recovery, Static Cramer and Scherbius Drives

The Slip Power Recovery Scheme

When rotor copper loss is neglected, P_r is equal to slip power, sP_g, where P_g is air gap power. Speed control below synchronous speed is obtained by controlling the slip power, without the use of rotor resistance control. Therefore, there methods of speed control are classified as slip power recovery scheme. Two such schemes:

Static Scherbius Drive

It provides the speed control of a wound rotor motor below synchronous speed. A portion of rotor AC power is converted into DC by a diode bridge. The controlled rectifier working as an inverter converts it back to AC and feeds it back to the AC source. Power fed back (i.e) P_r can be controlled by controlling inverter counter emf V_{d2}, which in turn is controlled by controlling the inverter firing angle. The DC link inductor is provided to reduce ripple in DC link current I_d.

Since slip power is fed back to the source, unlike rotor resistance control where it is wasted in resistors, drive has a high efficiency. The drive has higher efficiency than stator voltage control by AC voltage controllers because of the same reasons.

Drive input power is the difference between motor input power and the power fed back. Reactive input power is the sum of motor and inverter reactive powers. Therefore, drive has a poor power factor throughout the range of its operation.

From the figure, neglecting stator and rotor drops we have:

$$V_{d1} = \frac{3\sqrt{6}}{\pi}\frac{SV}{n} \quad ...(1)$$

$$V_{d2} = \frac{3\sqrt{6}}{\pi}\frac{V}{m}\cos\alpha \quad ...(2)$$

Where,

α - The inverter firing angle.

n and m - The stator to rotor turns ratio of motor and source side to converter side turns ratio of the transformer.

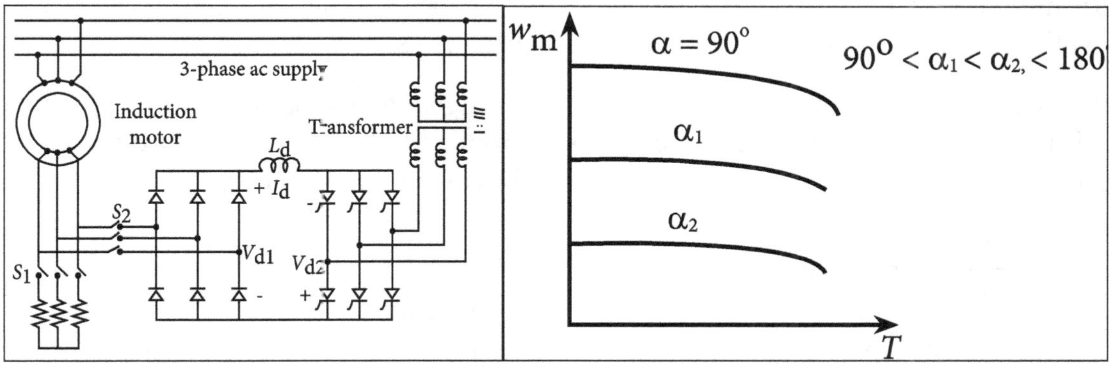

Static Scherbius Drive: (a) The drive circuit (b) Speed-torque characteristics.

Neglecting the drop across inductor, we have:

$$V_{d1} + V_{d2} = 0 \qquad \text{...(3)}$$

Substituting (1) and (2) in (3).

$$S = -\frac{n}{m}\cos\alpha = -a\cos\alpha$$

Where, a = n/m

Maximum value of a is restricted to 165° for safe commutation of inverter thyristors. Slip can be controlled from 0 to 0.966 α when α is changed from 90 to 165°. By appropriate choice of α, required speed range can be obtained.

Transformer is used to match the voltages V_{d1} and V_{d2}. At the lowest speed required from the drive, V_{d1} will have the maximum value V_{d1m} given by:

$Vd_{1m} = VS_{max}/n$ where S_{max} is the value of slip at the lowest speed. If α is restricted to 165°, m is chosen such that the inverter voltage has a value Vd_{1m} where α is 165°.

$$\frac{V}{m}\cos 165^\circ + \frac{V_{Smax}}{n}$$

$$m = -\frac{n\cos 165^\circ}{S_{max}} = -0.966\frac{n}{S_{max}}.$$

Such a choice of m ensures inverter operation at the highest firing angle at the lowest motor speed, giving highest power factor and lowest reactive power at the lowest speed. This improves the drive power factor and reduces reactive power at all speeds in the speed range of the drive.

Static Kramer Drive

The static Kramer-drive converts the slip power of an induction motor into AC power and supply back to the line. The slip power is the air gap power between the stator and the rotor of an induction motor which is not converted into mechanical power. Thus, the power is getting wasted. The static Kramer drives fed back the wasted power into the main supply. This method is only applicable when the speed of the drive is less than the synchronous speed.

The rotor slip power is converted into DC by a diode bridge. The DC power is now fed to DC motor mechanically coupled to induction motor. Torque supplied to load is sum of torque produced by induction and DC motor. Speed control is obtained by controlling field current of DC motor.

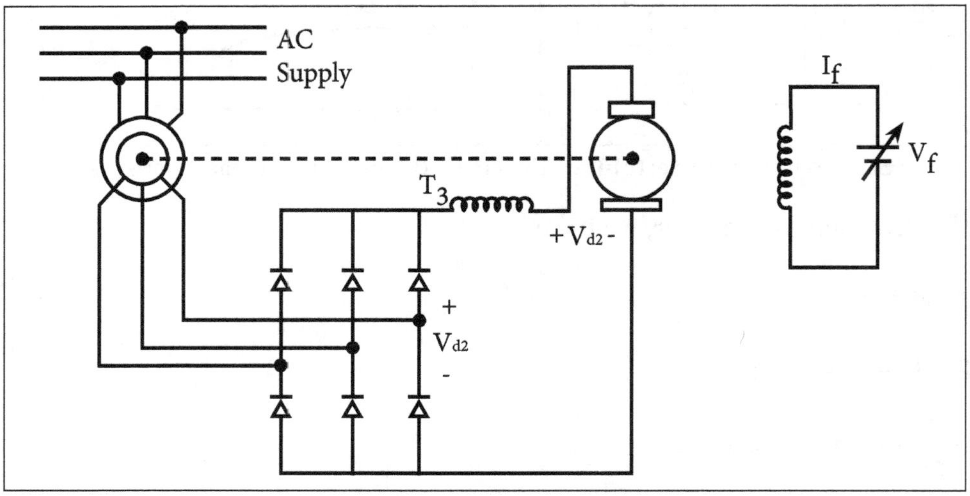

Static Kramer Drive Circuit.

Permissions

All chapters in this book are published with permission under the Creative Commons Attribution Share Alike License or equivalent. Every chapter published in this book has been scrutinized by our experts. Their significance has been extensively debated. The topics covered herein carry significant information for a comprehensive understanding. They may even be implemented as practical applications or may be referred to as a beginning point for further studies.

We would like to thank the editorial team for lending their expertise to make the book truly unique. They have played a crucial role in the development of this book. Without their invaluable contributions this book wouldn't have been possible. They have made vital efforts to compile up to date information on the varied aspects of this subject to make this book a valuable addition to the collection of many professionals and students.

This book was conceptualized with the vision of imparting up-to-date and integrated information in this field. To ensure the same, a matchless editorial board was set up. Every individual on the board went through rigorous rounds of assessment to prove their worth. After which they invested a large part of their time researching and compiling the most relevant data for our readers.

The editorial board has been involved in producing this book since its inception. They have spent rigorous hours researching and exploring the diverse topics which have resulted in the successful publishing of this book. They have passed on their knowledge of decades through this book. To expedite this challenging task, the publisher supported the team at every step. A small team of assistant editors was also appointed to further simplify the editing procedure and attain best results for the readers.

Apart from the editorial board, the designing team has also invested a significant amount of their time in understanding the subject and creating the most relevant covers. They scrutinized every image to scout for the most suitable representation of the subject and create an appropriate cover for the book.

The publishing team has been an ardent support to the editorial, designing and production team. Their endless efforts to recruit the best for this project, has resulted in the accomplishment of this book. They are a veteran in the field of academics and their pool of knowledge is as vast as their experience in printing. Their expertise and guidance has proved useful at every step. Their uncompromising quality standards have made this book an exceptional effort. Their encouragement from time to time has been an inspiration for everyone.

The publisher and the editorial board hope that this book will prove to be a valuable piece of knowledge for students, practitioners and scholars across the globe.

Index

www.ingramcontent.com/pod-product-compliance
Lightning Source LLC
LaVergne TN
LVHW080917210926
848485LV00004B/670

* 9 7 8 1 6 4 7 2 5 4 3 2 2 *